Horizontal Well Technology

Horizontal Well Technology

Editor

Rakesh Srivastav

Horizontal Well Technology
Edited by **Rakesh Srivastav**

Printed in 2017

ISBN: 978-1-68117-400-6

Library of Congress Control Number: 2015941591

Notice

Contents

Preface

Horizontal well technology was originally developed for use in petroleum production and underground utility installation, but recently has been adapted for environmental remediation applications. In the environmental remediation industry, horizontal wells provide unique characteristics and advantages that can improve the effectiveness of established soil and groundwater clean-up technologies now using traditional vertical well techniques. To date, over 300 horizontal wells are estimated to have been installed for environmental remediation purposes. This book was difficult because of the interdisciplinary nature of horizontal well technology. The book is mainly directed to the practicing professionals who make engineering calculations and decisions on horizontal well applications. This book can also be used as a graduate level textbook. For managers, the book helps to review the present state of the art.

Editor

Influence of the Chemical Composition of Completion Fluids on the Propagation of Electromagnetic Waves within Oil Wells

Alexandre Ashade Lassance Cunha,
Marco Aurélio Pacheco, and José Ricardo
Bergmann

Departamento de Engenharia Elétrica, Pontifícia Universidade
Católica do Rio de Janeiro, Rio de Janeiro, Brazil

ABSTRACT

The propagation of electromagnetic waves in the annular region of oil wells was studied. The present study aims to analyse the propagation attenuation along the well, as well as the input impedance determined by a source placed near the wellhead.

A coaxial waveguide model was adopted with heterogeneous dielectrics and losses. First, a wave equation solution for the waveguide is presented, assuming a homogeneous medium with losses, by solving the equation in cylindrical coordinates using the vector potential technique. An uncertainty analysis model is then developed to model the heterogeneous characteristics of the medium. Monte Carlo simulations were performed with the created model using data gathered from the literature. The results of the simulations indicate that propagation in the transverse electromagnetic mode has the smallest attenuation and that for depths of up to 4000 m, there is an attenuation of less than 52 dB. Furthermore, the input impedance ranges from 10 Ω to 10 kΩ because of the uncertainties involved in the problem in question.

INTRODUCTION

Oil wells today are extremely complex and demand very expensive maintenance [1]. Modern drilling techniques can reach a few kilometres in depth, and the costs of a floating platform at open sea can cost billions of Brazilian Reais. Thus, it is critical to determine the internal conditions of the well, and indicators such as temperature, pressure and salinity must be constantly monitored [2].

Reliable telemetry by cables is very difficult to obtain due to the extreme conditions of the internal environment of an oil extraction well. As an example, there is continuous abrasion by sand and dirt that are carried by fluid flow. For this reason, the cables are periodically damaged and require replacements. Such replacements hinder the oil extraction process and, moreover, increase the costs due to the large amount of cables needed over the life of a single well. The most obvious alternative to cabling is the adoption of wireless telemetry. However, the amount of electricity required to power wireless communication is not practical: power cables are needed because the use of batteries would lead to frequent maintenance, as they need to be recharged [1]. It is obvious, therefore, that the objective (and greater challenge) is to build a system without wires or batteries; that is, the downhole sensors must be powered and communicate with the base without the use of cables. Several approaches are possible for wireless telemetry. One possibility would be the use of signal transmission techniques using a magnetic field [3]. In this method, a coil that is capable of inducing alternating current through the production pipe is used. This coil has sufficient intensity to transmit both power and the signal itself between the sensor and the base [4]. Through a second coil positioned at the sensor location, it is possible to recover the signal and power needed to feed and produce bilateral communication. However, the technique has a weakness: the production tube is not a transformer core, i.e., it is not designed to minimise the magnetic flux losses. Its hollow characteristic confers much loss by parasitic Foucault currents concomitantly with hysteresis losses, the cause of which is due to the inadequacy of the production tube material for magnetic purposes [5]. Therefore, for long distances, on the order of 3000 m or more, the method is unfeasible.

Another possible approach would be to analyse the oil well from the perspective of a coaxial propagation structure, which is formed by a conductor tube with a perfect centre (production tube) and a cylindrical shell that is coaxial with the tube and is also a perfect conductor. This approach makes it possible to conduct the analysis as it would be performed on a coaxial cable, and therefore, propagation occurs in the transverse electromagnetic mode [6]. This approach, however, has been studied without considering fluids of significant conductivity or possible uncertainties caused by temperature and electrical parameter variations of the fluid that fills the well. The modelling difficulty concerns the fluid that fills the well. This fluid is heterogeneous because it exhibits significant variation in electric permittivity along the well depth. Furthermore, the actual fluid concentration varies from well to well, and at times within the well itself, which is another source of heterogeneity.

In this context, the present article proposes to study the electromagnetic propagation inside the annular area of oil wells. The goal is to develop a model that permits an approximation of the behaviour of the TEM propagation mode in the coaxial structure with losses and the previously cited heterogeneous characteristics.

The first step is the deduction of the electric and magnetic field equations within the well, assuming a typical well structure. Thereafter, statistical models of the parameters that compose the medium are developed, considering their temperature variations and propagation frequency. Finally, the study uses Monte Carlo analysis to illustrate how the propagation attenuation behaves and to analyse the input impedance of the coaxial waveguide as a function of position.

MODELLING

Well Definition

The well will be modelled according to Figure 1. This figure depicts a coaxial guide bounded by perfect conductor metals with a homogeneous dielectric and losses. The upper extremity is completely closed by a perfect conductor metal to model the valve and metal duct assembly present in a wellhead. The bottom of the well is composed of concrete; however, this fact is neglected, and the well is assumed to behave as a semi-infinite waveguide.

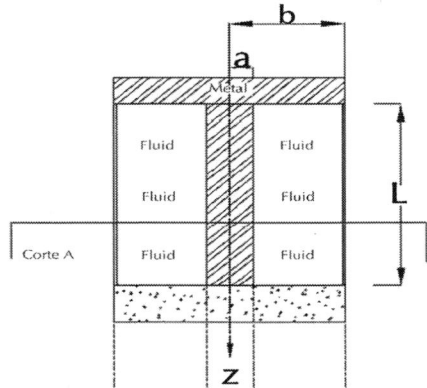

Figure 1: Adopted well structure. A semi-infinite well is assumed and is

formed by two concentric cylinders and a metal block covering the top of the waveguide.

Deduction of the Propagation Equations

Modelling is performed using Maxwell's equations and the concept of vector potentials. The Maxwell's equations used are Equations (1) and (2):

$$\nabla \times E = -M_i - j\omega B \tag{1}$$

$$\nabla \times H = J_i + \sigma_e E + j\omega\varepsilon' E \tag{2}$$

The real number constant ε' is the real part of the complex electrical permittivity of the medium. Furthermore, the complex constant σ_e is called effective conductivity and it represents the linear relation of conduction current and electric field on the medium.

Consider a medium with no source of magnetic charge $(\nabla.B = 0)$. Because $(\nabla.\nabla\times = 0)$, the curl of A can be defined as

$$\nabla \times A \equiv B = \mu H \tag{3}$$

Substituting 3 in 1 and considering $M_i = 0$ (Induced magnetic flux equal to zero), we obtain:

$$\nabla \times E = -\nabla \times j\omega A \tag{4}$$

Or:

$$\nabla \times (E + j\omega A) = 0 \tag{5}$$

Since $(\nabla \times \nabla = 0)$, an electric scalar potential g_e can be defined such that:

$$E + j\omega A \equiv -\nabla g_e$$

$$(6)$$

Applying the identity $(\nabla \times \nabla \times = \nabla(\nabla.) - \nabla^2)$ in Equation (3) and using Equation (2), we obtain:

$$-\nabla\nabla \cdot A + \nabla^2 A$$

$$= -\mu J_i + \mu(\sigma_e + j\omega\varepsilon')(\nabla g_e + j\omega A)$$

$$\sigma_e \equiv \omega\varepsilon'' + \sigma_s$$

$$(7)$$

In Equation (7), the parameter σ_e is the effective conductivity of the medium, while σ_s represents the static conductivity, that is, the conductivity when frequency is zero. At this moment, we are in a position to define $\nabla.A$. To simplify Equation (7), we use the gauge:

$$\nabla \cdot A \equiv -\frac{\gamma^2}{j\omega} g_e$$

$$(8)$$

where $\gamma^2_1 \equiv j\omega\mu(\sigma_e + j\omega\varepsilon')$ is the complex constant of propagation of the medium. This step simplifies the equation for the electrical potential, resulting in:

$$\nabla^2 A = -\mu J_i + \gamma^2 A$$

$$(9)$$

For homogeneous media, Equation (9), when solved, determines the electric vector potential in the medium, which can be used to obtain the equation for the electric field as a function of the electric vector potential:

$$E = \frac{j\omega}{\gamma^2}\nabla\nabla \cdot A - j\omega A$$

$$(10)$$

We seek the solution in transverse electromagnetic mode (TEM), which is generated using $A = \hat{a}_z A_z$ with the restriction $E_z = 0$, which leads to the following fields in cylindrical coordinates:

$$E_\rho = \frac{j\omega}{\gamma_1}\frac{A}{\rho}\sinh(\gamma_1 z)$$

$$(11)$$

$$H_\phi = -\frac{1}{\mu}\frac{A}{\rho}\cosh(\gamma_1 z)$$

$$(12)$$

$$\gamma_1^2 \equiv j\omega\mu(\sigma_e + j\omega\varepsilon')$$

$$(13)$$

Note that because $E_\Phi = E_z = 0$, the boundary conditions that require a null tangential component in the metal extremities are already met. As the well has a metal structure at one of the extremities, one must also ensure that E_ρ (z = 0) = 0. Note, however, that this condition is also already guaranteed.

The above solution represents a propagation model in TEM mode in a coaxial medium with losses inherent to the medium inside. However, it is important to remember that the solution was obtained by assuming a homogeneous propagation medium.

Statistical Modelling of the Medium Constituent Parameters

The propagation medium in the well consists of an oilbased dielectric fluid composed of water, oil and salts (typically $CaCl_2$) [7]. This fluid is the centre of all electromagnetic propagation, and therefore, its electrical characteristics must be examined in detail. A study over the range of 1 MHz to 100 MHz was previously conducted [7], revealing significantly variable behaviour based on the chemical composition of the fluid.

To model the variation of conductivity with frequency, the effective conductivity concept is applied [8] using the following formula:

$$\sigma_e \equiv \sigma_s + \omega\varepsilon''$$

$$(14)$$

From the experimental curves obtained in [8], the parameters σ_s and ε'' can be calculated using a least squares method.

An analogous model can be created to model the variation in frequency of the real relative permittivity with the frequency. With

$$\varepsilon' \equiv \varepsilon_s + K\omega$$

(15)

and using the experimental curves in [8], it is possible to estimate the parameters and K using a least squares approach.

In addition to the variation in frequency, variation in the medium ε_s constituent parameter can also be observed in temperature. Such variation cannot be neglected because inside a 5000 m deep well, it is impossible to ensure that the temperature is uniform along its entire length. This fact, therefore, characterises non-homogeneity along the z direction.

To circumvent the situation, a model that assumes an average temperature in the medium is adopted. This temperature, in turn, is considered constant throughout the well, which implies a homogeneous medium. Thus, the influence of the variation of this average temperature on signal propagation in the well can be analysed, assuming a valid range of average temperatures.

Mathematically, a coefficient of correction in temperature is defined as the ratio between the temperature value in question and the value of the reference temperature, here defined as 25°C. Thus,

$$\theta \equiv \frac{\sigma_e(\theta)}{\sigma_e(25°C)}$$

(16)

$$\theta \approx a_1\theta^2 + a_2\theta + a_3$$

(17)

$$\sigma_e = \sigma_e(25°C) \times \theta$$

(18)

The quadratic form was selected to present the best interpolating results using the experimental data from [8].

Similarly, there is increasing variation in the relative permittivity according to the temperature, as demonstrated by [7]. Again, it is used a quadratic model similar to the effective conductivity variation in temperature:

$$\theta_\varepsilon \equiv \frac{\varepsilon'(\theta)}{\varepsilon'(25\,^\circ C)}$$

(19)

$$\theta_\varepsilon \approx a_1\theta^2 + a_2\theta + a_3$$

(20)

$$\varepsilon' = \varepsilon'(25\,^\circ C) \times \theta_\varepsilon$$

(21)

EXPERIMENTS AND RESULTS

In all the experiments, a well with a length of 5000 m, an inner radius of 0.05 m and an outer radius of 0.1 m is assumed. All the Monte Carlo analyses were conducted with at least 1 million samples. The objective of the experiments was to obtain graphs and numerical values for the input impedance of the "oil well" waveguide and to analyse the propagation loss in the medium for 5000 m of depth.

All analyses were conducted over the range of 1 MHz to 100 MHz. For the other free parameters, the modelling used random variables whose distribution was selected to reflect their most common values. Table 1 summarises the values selected for each free parameter of the previously described model and their respective distributions. For simplicity, when the random variable has simple and obvious domain restrictions, a uniform distribution was selected; otherwise, a normal distribution was used. Furthermore, statistical independence was assumed between the variables.

First, an experiment was conducted to calculate the attenuation in the well. The attenuation is directly dependent on the real part of the constant of propagation, whose square is defined as

$$\gamma^2_1 \equiv j\omega\mu(\sigma_e + j\omega\varepsilon').$$

The models proposed in Section 2.3 were used for the conductivity and electric permittivity, and the relative magnetic permeability was assumed to be equal to 1.

The Figure 2 presents a boxplot of the constant of attenuation for the various frequencies between 1 and 100 MHz. Values that appear in the graph as outliers are from highly unlikely combinations of the random variables of the problem and are most likely not physically feasible and, therefore, must be disregarded. The height of the boxes represents the range of values most likely to be observed, and its average point is a good approximation of the average. The boxplot of Figure 3demonstrates that the attenuation constant increases with an increase in frequency, as expected. Furthermore, note that the deviation of the coefficient increases with the elevation of the propagation frequency, which makes the system design even more difficult. Thus, it is clear that lower frequentcies are desirable from the point of view of signal attenuation.

Table 1: Random variables used for modelling uncertainties. For simplicity, when the random variable has simple and obvious domain restrictions, a uniform distribution was selected; otherwise, a normal distribution was used

Variable	Average/a	Deviation/b	Distribution
σ_s (S/m)	0	4.0×10^{-5}	Uniform
ε_s (F/m)	3	12	Uniform
ε'' (F/m)	1.0×10^{-13}	3.0×10^{-12}	Uniform
$K(F/H_z\ m)$	2.5×10^{-9}	5.0×10^{-10}	Normal
$\theta_{med}\ °C$	40	5	Normal
a_1(permissivity)	9.0×10^{-4}	2.0×10^{-4}	Normal
a_2(permissivity)	-3.0×10^{-12}	6.0×10^{-3}	Normal
a_3(permissivity)	1.2	0.1	Normal
a_1(permissivity)	1.0×10^{-5}	2.0×10^{-6}	Normal
a_2(permissivity)	0.0	5.0×10^{-4}	Normal
a_3(permissivity)	1.0	1.0×10^{-2}	Normal

Using the TEM mode equations presented in Equation (2), it is clear that the wave propagating in the direction of the bottom of the well has power attenuation given by $A = \exp(\alpha^2 L^2)$. Using the statistical analysis of the coefficient of attenuation, Table 2 was generated.

The values from the table represent the upper limit of attenuation for 95% of the cases. Thus, for example, for a depth of 1000 m, P ($A \leq 13$ dB) = 0.95. The results presented in Table 2 do not agree with the propagation for L = 5000 m of depth because of the high-energy attenuation (130.3 dB). However, it is essential to note that the table represents an upper limit of attenuation, taking 95% of the possible combinations of propagation medium and an average temperature. Moreover, as the table demonstrates, propagations up to 2000 m depth are highly acceptable, as an attenuation of up to 52.1 dB is observed in current communication technology. Even at greater depths, the possibility of communication cannot be excluded.

The second experiment was performed to analyse the input impedance of the coaxial waveguide as a function of the excitation source position. The experiment was conducted at 1 MHz.

Setting the impedance at any point as the ratio E_ρ / H_ϕ, we obtain

$$Z = Z_0 \tanh\left(\gamma_1 z\right) \tag{22}$$

$$-L < z \leq 0 \tag{23}$$

$$Z_0 = -\mu \frac{j\omega}{\gamma_1} \tag{24}$$

The position selected for analysis was at 1/4 wavelength from the wellhead, that is, $Z = \lambda / 4$. However, the wavelength itself is a random variable because $\lambda = 2 * \pi / \beta$ and $\beta = \text{Im}\{\gamma\}$ is a random variable. Monte Carlo analysis revealed that γ varies between 90 m and 170 m for 90% of the cases, leading to the selection of position $Z = ((90 + 170) * 1/2)/4 = 32.5\text{m}$.

Table 2: Estimated attenuation of electromagnectic waves propagating in the annular region of the well

L(m)	Electric field attenuation (dB)	Power attenuation (dB)
1000	13.0	26.1
2000	26.1	52.1
3000	39.1	78.2
4000	52.1	104.2
5000	65.1	130.3
6000	78.2	156.3
7000	91.2	182.4

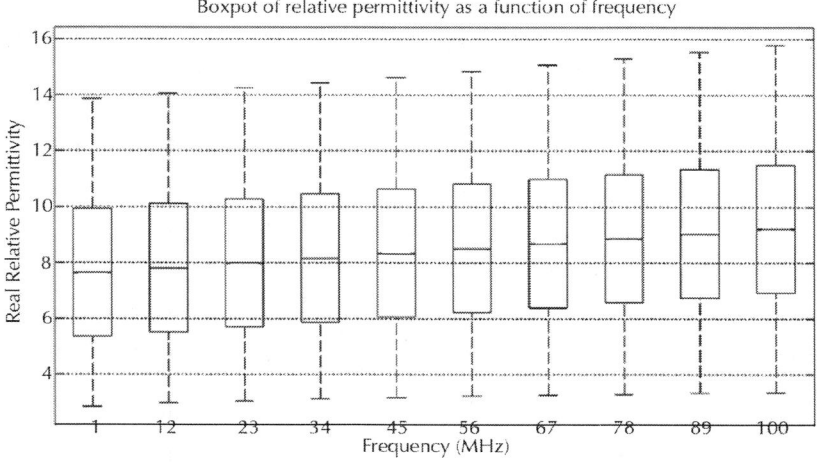

Figure 2: Boxplot of relative permittivity as a function of frequency. The horizontal axis represent the frequency of propagation in MHz, while the vertical axis represent the actual value of the electrical permittivity.

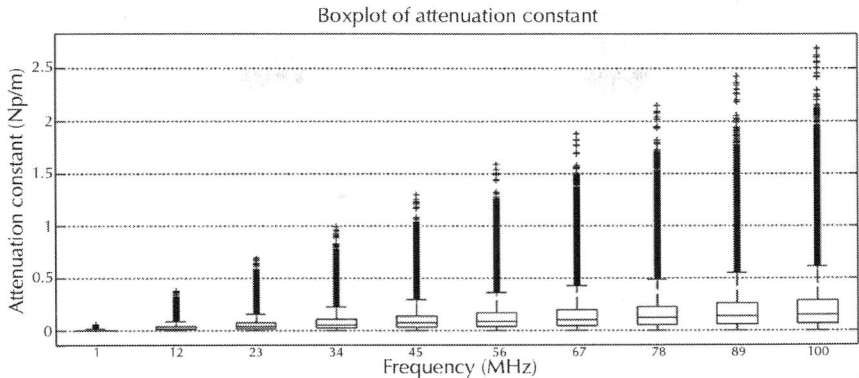

Figure 3: Boxplot of attenuation constant as a function of frequency.

Using Monte Carlo analysis again, it can be observed in Figure 4 that the input resistance (real part of the impedance) of the system exhibits great variation, ranging from 10 Ω to 1.0 kΩ for 90% of the cases. This variation is due to the variation of the relative permittivity in the medium. Therefore, it is necessary to design a generator circuit that provides a good match for a wide range of input impedances, i.e., the generator/receiver circuit must lose as little power as possible by reflection.

CONCLUSIONS

The present study focused on analysis of the electromagnetic propagation inside the annular area of oil wells, assuming the interior medium is composed of dielectrics with significant conductivity. The well behaviour was evaluated with respect to the input impedance and propagation attenuation.

To quantify the well behaviour with respect to electromagnetic propagation, a coaxial waveguide model was developed, modelling the wellhead as a metal cap that completely closes one extremity of the waveguide. First, we solved the wave equation for a homogeneous cylindrical coaxial waveguide, resulting in an analytical model. Then, an uncertainty model was adopted to approximate the heterogeneous and imprecise characteristics of the fluid that functions as a dielectric inside the well.

Using this model, two experiments were developed, both by simulation. The first aimed at analysing the input impedance observed by a source positioned at 1/4 of the wavelength from the wellhead, while the other aimed at analysing the attenuation as a function of length or well depth.

Due to the uncertainties present, the input impedance was observed to vary from 10 Ω to 1 kΩ for the frequency of 1 MHz. The conclusion was obtained by Monte Carlo simulation and was applied to the expression derived for the waveguide input impedance.

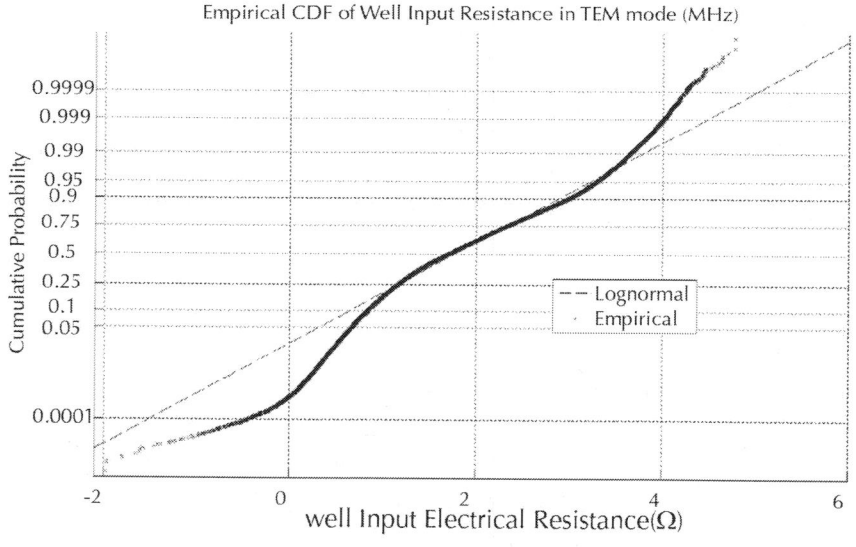

Figure 4: Empirical cumulative distribution function of the well input resistance in TEM mode at 1/4 wavelength from well head. The horizontal axis show the log on base 10 of the value of input resistance in Ohms, while the vertical axis show the cumulative distribution function value.

By performing the attenuation analysis, it was concluded that for 95% of the cases, the constant of attenuation in TEM mode is less than 0.8×10^{-4} Np/m for a frequency of 1 MHz. The power attenuation at 4000 m depth was also observed to be approximately 100 dB for the same frequency of 1 MHz, with 95% probability. Although 100 dB appears to be a large attenuation, most wells in operation are between 1000 m and 2000 m in length, and the attenuation is much lower at these depths.

REFERENCES

1. S. Brilles, "Remote Downhole Well Telemetry," US Patent US6766141, 2004.

2. J. A. D. Rosa, A. J. Carvalho and R. D. S. Xavier, "Petroleum Reservoir Engineering," Rio de Janeiro, 2006.

3. F. Sakata, H. Wakiwaka, M. Hanabusa, N. Yamazaki and H. Yamada, "Performance Analysis of Long Distance Transmitting of a Magnetic Signal in a Cylindrical Steel Rod," IEEE Translation Journal on Magnectics in Japan, Vol. 8, No. 2, 1993, pp. 102-106.

4. F. Harold J. Vinegar, R. R. Burnett, G. C. W. M. Savage and J. W. Hall, "Permanent Downhole, Wireless, TwoWay Telemetry Backbone Using Redundant Repeaters," US Patent US6633236B2, 2003.

5. B. W. Kennedy, "Energy Efficient Transformers," New York, 1998.

6. K. A. Safynia and R. W. McBride, "System and Method for Communicating Signals in a Cased Borehole with Tubing," US Patent US4839644, 1989.

7. P. A. Patil, et al., "Experimental Study of Electrical Properties of Oil-Based Mud in the Frequency Range from 1 to 100 MHz," SPE Drilling and Completion, Vol. 25, No. 3, 2010, pp. 380-390. doi:10.2118/118802-PA

8. C. A. Balanis, "Advanced Engineering Electromagnetics, Vol. 52, No. 1," Wiley, Hoboken, 1989, p. 1008.

Chapter 2

Effect of Horizontal and Vertical Well Patterns on Methane Hydrate Dissociation Behaviors in Pilot-scale Hydrate Simulator

Jing-Chun Feng[a, b, c], Yi Wang[a, b], Xiao-Sen Li[a, b],
Gang Li[a, b], Yu Zhang[a, b], and Zhao-Yang Chen[a,b]

[a]Key Laboratory of Gas Hydrate, Guangzhou Institute of Energy Conversion, Chinese Academy of Sciences, Guangzhou 510640, PR China

[b]Guangzhou Center for Gas Hydrate Research, Chinese Academy of Sciences, Guangzhou 510640, PR China

[c]University of Chinese Academy of Sciences, Beijing 100083, PR China

ABSTRACT

Exploitation of natural gas hydrate is expecting to be an important strategic way to solve the problem of energy depletion. Understanding the effectiveness of the well configuration plays a pivotal role in gas production from the hydrate reservoir. This study evaluates the methane hydrate dissociation behaviors using both vertical well and horizontal well experimentally. Methane hydrate in porous media has been synthesized in a 117.8 L pilot-scale hydrate simulator (PHS), which is equipped with 9 (3 × 3) vertical wells and 9 (3 × 3) horizontal wells. The condition of hydrate formation is corresponding to the ocean depth of 1200 m and it is similar to the hydrate characteristics of the South China Sea. Hydrate is dissociated under depressurization and thermal stimulation. The results indicate that, for the depressurization and thermal stimulation methods, the gas production rate, the heat transfer rate, and the accumulative dissociation ratio with the horizontal well pattern are higher than those with the vertical well pattern. Meanwhile, the evaluations of the energy ratio and the thermal efficiency indicate that the horizontal well pattern has the advantage of higher production efficiency by the thermal stimulation. Thus, it is determined that the production performance is better using the horizontal well pattern.

INTRODUCTION

Natural gas hydrates are crystalline solid compounds which are formed at high pressures and low temperatures in the permafrost region and subsea floor [1]. Although it is uncertain, the common estimation of the global methane hydrate is as large as 2×10^{16} m^3 on the continental margins [2] and [3]. Gas hydrate has been considered as a potential strategic energy on account of the abundant carbon gas trapped in this resource [4]. At present, common methods proposed for hydrate dissociation are depressurization [5], [6],[7], [8] and [9], thermal stimulation [10], [11] and [12], inhibitor stimulation [13], [14] and [15], carbon dioxide replacement [16] and [17], and the combined utilization of these methods [18]. In addition to the dissociation methods, the gas production is highly influenced by the well patterns, because the heat-transfer and the fluid flow in the hydrate reservoir are controlled by the well configuration. The configuration of well pattern not only should be according to the intrinsic character of the hydrate reservoir, but also is restricted to the practical geological conditions. Some researchers have investigated this issue by the numerical simulations. For example, Moridis et al. [19] indicated that horizontal well was impractical at the UBGH2-6 site of the Ulleung basin in the Korean East Sea due to the layered stratigraphy and the presence of mud layers. Their investigation indicated that production from such a hydrate accumulation with the vertical well was feasible. While in the Qilian Mountain permafrost, the numerical simulation of Zhao et al. [20] showed that the single vertical well by depressurization was not promising, and they suggested that other methods such as a single horizontal well design may be more economical for gas production at the Qilian Mountain permafrost. Li et al. [21] investigated the gas production performance by the huff and puff method with a single horizontal well in the Shenhu Area, and the results showed that the gas production rate was lower than the acceptable standard of commercial gas production. The similar result was carried out by Su et al. [22] when the numerical simulation was conducted by the huff and puff method with a single vertical well. Furthermore, Feng et al. [23] reported that in the South China Sea, the dual horizontal wells placed in the same horizontal plane has the advantage of higher productivity than that placed in the same vertical plane.

Not only the numerical simulation, but also the field test and the laboratory test are essential to investigate the feasible exploitation technology of methane hydrate. So far, because the field test is costly and time-consuming, only Japan has carried out the field test of marine hydrate exploitation in the Nankai Trough in May 2013 [24]. Consequently, the laboratory test acts as a significant role in the study of gas production technology. Recently, the different sizes of the large-scale reactors have been built to simulate the formation and dissociation processes of the hydrate. A 72.2 L seafloor process simulator [25] has been manufactured at the Oak Ridge National Laboratory, USA. This reactor has been applied for the investigation into the hydrate formation and dissociation under the condition of marine water depth of 2 km. Later on, a 72 L reactor was fabricated by Zhou et al. [26] to investigate the methane gas production behaviors from methane hydrate by depressurization. Fitzgerald et al. [27] have investigated the net energy efficiency of gas production by in-situ heating in a 70 L reactor. The results showed that the net energy efficiency of 72% were achieved during 10 h test period with the heating at 100 W. In addition, within the framework of the German national research project, a 425 L large laboratory reservoir simulator (LARS) was set up by Schicks et al. [28] to study the thermal stimulation by using in situ combustion. They also investigated the CH_4–CO_2 swapping process in this simulator. Moreover, a High-pressure Giant Unit for Methane-hydrate Analyses (HiGUMA) was developed by Kono et al. [29] to study the gas recovery factor of methane hydrate by different schemes of depressurization in sandy sediment. The internal volume of the HiGUMA is 1710 L, furnished with a single vertical system. These large-scale hydrate simulators have made great contributions to studying the formation and dissociation conditions of the hydrate as well as the feasible exploitation technology for the future.

However, all of the above-mentioned hydrate simulators were only equipped with a single production well. It was hard to investigate the different well patterns on gas production from hydrate bearing sediment by using these simulators. Li et al. [30] built a 117.8 L cylindrical hydrate simulator, which was equipped with 9 (3 × 3) vertical wells and 9 (3 × 3) horizontal wells, making it the unique hydrate simulator associated with multi-well configurations. This pilot-scale hydrate simulator (PHS) consists 147 (7 × 7 × 3) temperature measuring points, making it possible to precisely predict the thermal front in the process of hydrate formation and dissociation. In recent years, they had assessed the production performance of methane hydrate in the sandy reservoir through a single vertical well by the depressurization and the huff and puff method [31]. They also introduced the SAGD method from oil industry into hydrate decomposition with the PHS, in which the dual horizontal wells were applied [32].

Up to now, little attention has been paid to the studies of the different well configurations in the laboratory tests. Few references reported about the comparison of the effects of the vertical well and the horizontal well on the hydrate dissociation behaviors in the porous media sediment by numerical or experimental simulation.

In this work, the investigation into the effects of the vertical and horizontal well patterns on gas production behaviors has been carried out in a pilot-scale hydrate simulator. The central vertical well and the central horizontal well are selected as the production well. In the practical field, the horizontal well can be much longer than the vertical well because most of the distributions of the hydrate reservoirs are horizontal layered. To compare the effect of the same well section, the vertical and horizontal wells are of the same length in this study. Meanwhile, the depressurization and the thermal stimulation are employed to dissociate the hydrate. The cumulative gas production, the accumulative dissociation ratio, the energy ratio and the thermal efficiency are selected as the indicators to evaluate the production performance.

EXPERIMENTAL SECTION

Experimental Apparatus

The detailed information of the PHS has been reported in the previous studies [6] and [31]. Fig. 1 shows a schematic drawing of the apparatus. A cylindrical pilot-scale hydrate simulator (PHS) fabricated from 316 stainless is the core of the apparatus. The maximum working pressure of the PHS is 30 MPa. The inner effective volume of the PHS is 117.8 L with the height of 600 mm, and the diameter of 500 mm. To avoid the non-uniform effects of boundary temperature, the PHS is placed in a cold room (−8–30 °C, ±30 °C) and it is surrounded by a water jacket (−15–30 °C, ±0.1

°C) to maintain the temperature stable. To measure the pressure of the system, an inlet and an outlet pressure transducer are placed in the top and bottom of the PHS, respectively. The production pressure at the well is adjusted by a back-pressure regulator situated in the outlet (from the Nantong FeiYu Company, 0–30 MPa, ±0.2 MPa). A metering pump (from Beijing ChuangXinTongHeng Company, 0–250 mL/min, ±0.1 mL/min) is acted as the injection equipment, and a heater (from the Nantong FeiYu Company, 0–6 kw, ±0.1 kw) is used to provide the hot water or steam. A gas/liquid separation equipment is placed in the outlet to separate the produced fluid. Two gas flow meters (from the Seven Star Company, 0–100 L/min, ±2%) are applied to measure the injected gas and produced gas, respectively. The quantity of the water production is measured by a balance (from the Guangzhou ZhiCheng Electronic Scale Company, 0–30 kg, ±1 g). All of the parametric variations during the experiment are collected by the data acquisition system.

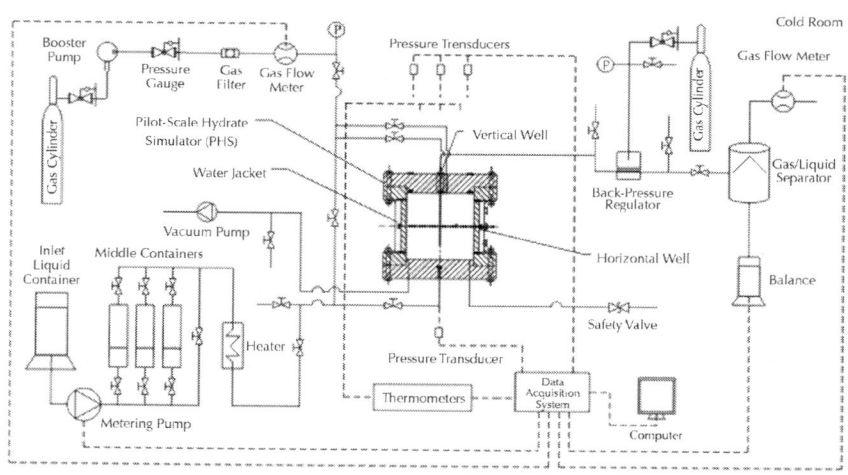

Figure 1: Schematic of the experimental apparatus.

Fig. 2 gives the schematic of the inner PHS and the well configurations in this work. The inner PHS is divided into four regions of the same size by three horizontal layers, which are named as Layers A–C, respectively. The central vertical well is situated along the axis of the PHS. The central horizontal well is placed in the middle of the Layer B. Both the gas production by the depressurization and the thermal stimulation are conducted with the central vertical well and the central horizontal well, respectively.

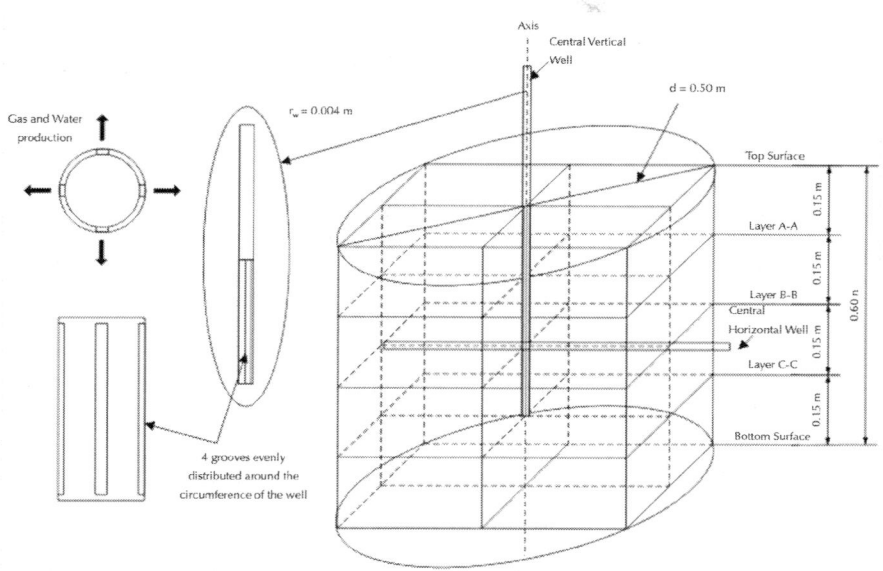

Figure 2: Schematic of the inner PHS and the well configuration.

Fig. 3 shows the schematic of the distribution of the thermal couples in the PHS. As seen in Fig. 3, there are 49 (7 × 7 = 49) thermal couples on each layer. Therefore, the total amount of the thermal couples in the PHS is 149 (49 × 7 = 147). The name of the thermal couples in the PHS can be explained as follows: as an example, the 43rd thermal couples situated on Layers A–C are named as T43A, T43B, and T43C, respectively.

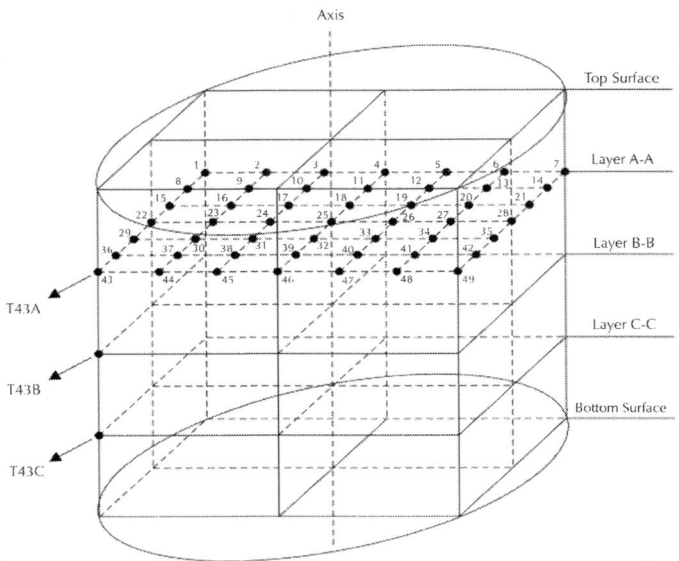

Figure 3: Schematic of the distributions of the thermal couples in the PHS.

Experimental Procedure

Hydrate Formation

Table 1 lists the formation and dissociation conditions for the experiments. The detailed information of the hydrate formation process has been introduced by Wang et al. [33]. Firstly, quartz sands (the particle size ranges from 300 to 450 μm) were tightly packed in the reactor, and the porosity of this sandy sediment was 43.5%. Then, a vacuum pump was used to empty the reactor and expel the residual gas. The precalculated amount of 22,433 g deionized water was injected into the PHS. Next, the reactor was pressurized to approximately 20 MPa by the methane gas. The amount of methane pumped into the reactor at this condition was 28,926 L. Hydrate gradually formed in the PHS with the decrease of the system pressure.

Table 1: Hydrate formation and dissociation conditions

Parameter	Run 1	Run 2	Run 3	Run 4
Well configuration	Vertical well	Horizontal well	Vertical well	Horizontal well
Production method	Depressurization	Depressurization	Thermal stimulation	Thermal stimulation
Initial pressure after hydrate formation (MPa)	13.55	13.49	13.55	13.49
Initial hydrate saturation	0.28	0.27	0.28	0.28
Initial gas saturation	0.51	0.51	0.51	0.52
Initial water saturation	0.21	0.22	0.21	0.20
Water injection rate (mL/min)	/	/	200	200
Temperature of the injected water (°C)	/	/	44.87	45.34

The targeted gas, water, and hydrate saturation (S_G, S_W, S_H) in this investigation were 50%, 20%, and 30%, respectively. The determination of the three-phase saturation for hydrate formation was calculated through the mass balance relationship as follows [6]:

$$S_G + S_W + S_H = 1 \tag{1}$$

$$S_G = \frac{v_m n_{m,G}}{V_{pore}} \tag{2}$$

$$S_W = \frac{m_{W,inj} - N_H(n_{m0} - n_{m,G} - n_{m,W})M_W}{\rho_W V_{pore}} \tag{3}$$

$$S_H = \frac{(n_{m0} - n_{m,G} - n_{m,W})M_H}{\rho_H V_{\text{pore}}}$$

(4)

where v_m represented the molar volume of methane gas (mL/mol), which was calculated by the fugacity model of Li et al. [34]. n_{m0} was the amount of the methane gas pumped into the PHS. $n_{m,G}$ and $n_{m,W}$ were the remaining gas in gas phase and the dissolved gas in aqueous in the PHS, respectively. $n_{W,inj}$ was the initial amount of water injected into the reactor. V_{pore} was the pore volume of the quartz sand sediment. As the quartz sand was regarded as incompressible, V_{pore} acted as constant in this investigation. M_W and M_H were the molar mass of the water and the hydrate, respectively. ρ_W and ρ_H represented the density of the water and the hydrate, respectively.

The experimental pressure and temperature conditions originated from the geophysical data of the hydrate accumulation in the South China Sea [35], where the water depth is 1200 m, and the strata thickness under the sea floor is 145 m. The local pressure and temperature are 13.5 MPa and 8 °C, respectively. In this experiment, when the system pressure decreased to 13.5 MPa, the hydrate saturation reached the target value.

Hydrate Dissociation

In this work, the depressurization and the thermal stimulation are employed to dissociate the hydrate. Each method is used with the vertical well and the horizontal well, respectively. As shown in Table 1, runs 1 and 2 are conducted by depressurization. The gas production process began when the outlet valve was opened. When the pressure in the PHS declined to 4.7 MPa (the target value), it was maintained at approximately 4.7 MPa by adjusting the back-pressure regulator until the end. The depressurizing rate, depressurizing range and the boundary condition were kept as the same for runs 1 and 2.

Runs 3 and 4 were carried out by the thermal stimulation. Hydrate was designed to be dissociated under the driving force of thermodynamics. Hence, the huff and puff method with the production pressure at 6.5 MPa (above the corresponding equilibrium pressure 5.68 MPa) was employed. The procedure was explained as follows: the target value of the back-pressure regulator was set as 6.5 MPa. When the system pressure decreased to 6.5 MPa, the huff and puff method process started. The huff and puff process was divided into three stages. The first stage was the injection stage, in which the warm water was injected into the hydrate reservoir for 20 min with the injection rate of 200 ml/min. The second stage was the soaking stage, which lasted 30 min. During this stage, the inlet and outlet valve were closed for heat diffusion in the hydrate reservoir. The third stage was the production stage, in which the outlet valve was opened for gas production and the pressure of the system was kept at 6.5 MPa by regulating the back-pressure regulator. When the system pressure decreases to 6.5 MPa, the next cycle of injection stage began. The gas production process can be terminated when there was little gas released from the hydrate reservoir.

RESULTS AND DISCUSSION

Depressurization

Pressure and Temperature

Fig. 4 shows the evolutions of the pressures and temperatures in the hydrate reservoir for runs 1 and 2. As illustrated from the change of the pressure in the hydrate reservoir, the hydrate dissociation process can be divided into 2 stages. The first stage is the pressure reduction stage, in which the pressure is decreased from point P_A (13.5 MPa, the initial state after hydrate formation) to point P_B (4.7 MPa, below the equilibrium pressure). The second stage is the constant-pressure production stage, in which the system pressure is maintained at approximately 4.7 MPa till the end.

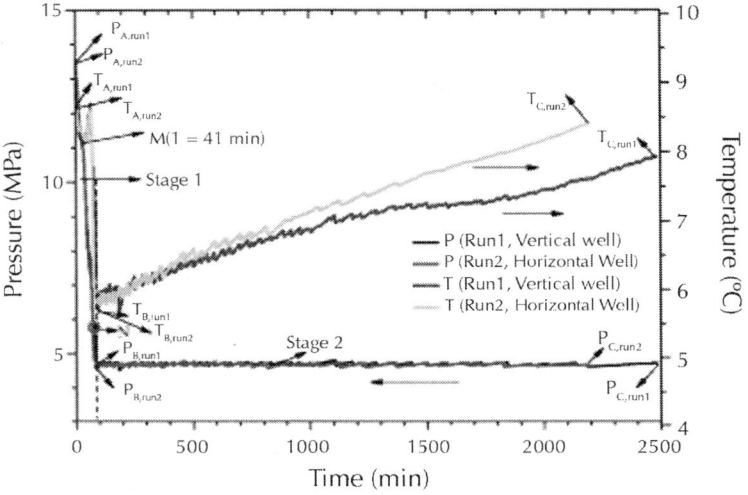

Figure 4: Evolution of pressure and average temperature in the PHS by depressurization.

During the first stage, run 1 lasts 85.67 min (from point $P_{A,run\ 1}$ to point $P_{B,run\ 1}$). The duration of run 2 is 83.67 min (from point $P_{A,run\ 2}$ to point $P_{B,run\ 2}$), which can be considered to be approximately identical to that of run 1. The depressurization amplitudes for the two runs are 8.8 MPa. Therefore, the depressurizing rate (depressurization amplitude/duration) can be considered to be the same (0.105 MPa/min) for the two runs. It is verified in Fig. 4 that the pressure curves for the two runs in these stages almost completely coincide. The temperature variation for run 1 is consistent with that for run 2. As shown in Fig. 4, before point M (t = 41 min) the temperatures decrease with the reduction of the system pressure caused by the Joule–Thomson effect [36]. Afterward, it is noted that there is a short-term temperature increase, which is caused by the heat release of hydrate re-formation. The reason for hydrate re-formation is that the pressure of the system at this time is above the equilibrium pressure, and the fluid flow in the porous media increases the contact interface area between water and gas dramatically, which is favorable for hydrate formation [37]. Subsequently, the system pressure gradually decreases to the equilibrium pressure (point N), and the hydrate starts to be dissociated. Meanwhile, the temperature is decreased drastically to the lowest point caused by the heat absorption of hydrate dissociation.

In the second stage, the pressures for the two runs both are kept at approximately 4.7 MPa. It is needed to absorb energy from the boundaries for hydrate dissociation, causing the increase of the average temperature. Fig. 4 shows that, for the horizontal well pattern, the temperature growth rate is faster than that for the vertical well pattern. This indicates that the heat transfer rate with the horizontal well pattern is faster than that with the vertical well pattern. The reason is that the heat convection rate is higher with the horizontal well. The heat convection rate is influenced by the mobility of gas and water in the sandy sediment. For the vertical well, the direction of gravity is the same with the axial direction of the well. However, the direction of gravity is perpendicular to the axial direction of the horizontal well. Therefore, the turbulent flow is more drastic with the horizontal well pattern under the effect of the gravity, causing the heat convection rate with the horizontal well pattern is larger than that with the vertical well pattern.

Gas and Water Production

Fig. 5 shows the evolution of the accumulative volume of gas production (V_p) for runs 1 and 2. In the first stage, the gas production curve for run 1 evolves in accordance with that for run 2, indicating that the rates of gas production for the two runs are equal. The reason is that the two runs are carried out under the same depressurizing rate (shown in Fig. 4). It is in good agreement with the conclusion of Li et al. [35] that the depressurizing rate is the dominant factor of gas production in the earlier procedure of hydrate dissociation by the depressurization. From point A to point N, all of the produced gas is the free gas remained in the vessel after hydrate formation, because the pressures are above the equilibrium pressure. The hydrate starts to be dissociated from point N. Hence, the produced gas in the later part of the first stage (from point N to point B) is the mixture of the free gas and the dissociated gas. All of the produced gas in the second stage (from point B to point C) is the dissociated gas because the pressure in this stage is kept below the equilibrium pressure.

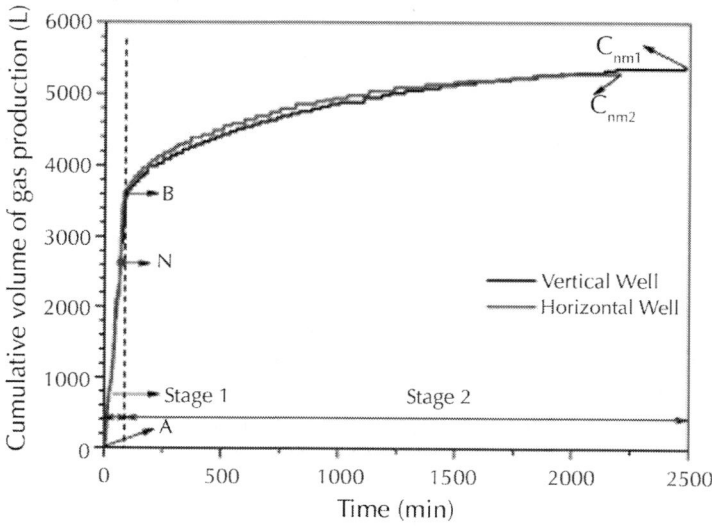

Figure 5: Evolution of cumulative volume of gas production by depressurization.

The production rates are mainly governed by the limitations of heat and mass transfers for the hydrate reservoir with a large scale [38]. As shown in Fig. 5, in the second stage, the total volume of the produced gas is almost identical for the two runs. However, the duration for run 2 (2113 min) is approximately 88% of the duration for run 1 (2397 min), indicating the hydrate is dissociated more quickly with the horizontal well pattern.

Accumulative dissociation ratio is defined as the ratio of the accumulative amount of the dissociated hydrate (m_{diss}) to the total amount of the hydrate at the initial moment (m_{hyd}), which can be expressed as follows [16]:

$$\varphi = \frac{m_{diss}}{m_{hyd}} \tag{5}$$

where m_{diss} means the amount of the dissociated hydrate, and m_{hyd} represents the initial amount of the hydrate in the PHS.

m_{diss} is calculated by the following equation:

$$m_{diss} = \rho_H (S_{H'0} - S_{H't}) \tag{6}$$

where $S_{H,0}$ is the initial hydrate saturation before hydrate dissociation. $S_{H,t}$ is the hydrate saturation at any given time during the dissociation process. For hydrae dissociation, the three phase saturation at any given time are explained as follows:

$$S_{H't} = S_{H'0}(1-\phi) \tag{7}$$

$$S_{G,t} = \frac{(n_{m,0} + S_{H,0}V_{pore}k\rho_H M_H - n_{m,w} - V_P/22.14)v_m}{V_{pore}} \tag{8}$$

$$S_{W,t} = S_{W,0} + \frac{N_H M_W \rho_H S_{H,0}\varphi}{M_H} - \frac{m_W}{\rho_W V_{Pore}} \tag{9}$$

where V_p and m_w are the accumulative amount of gas (L) and water production (g) during the process of hydrate dissociation. $n_{m,0}$ is the total amount of free gas (mol). The accumulative dissociation ratio can be calculated by combining Eq. (1) and Eqs. (5), (6), (7), (8) and (9).

Fig. 6 gives the evolution of the accumulative dissociation ratios for runs 1 and 2. As seen from Fig. 6, the changes of the accumulative dissociation ratios for runs 1 and 2 in the first stage are almost the same. The accumulative dissociation ratios are lower than 20% for both runs 1 and 2 at the end of the first stage, indicating that the majority of the hydrate is dissociated in the constant-pressure production stage. During the second stage, the accumulative dissociation ratios for the two runs gradually increase with time. The accumulative dissociation ratio for run 2 is higher at the fixed time than that for run 1, which is because the heat convection rate with the horizontal well pattern is stronger than that with the vertical well pattern. When the accumulative dissociation ratios for runs 1 and 2 increase up to 94.18% and 94.07%, respectively, the hydrate dissociation rates for runs 1 and 2 are quite low. Afterward, it means that, dissociating completely the remaining hydrate of less than 6% in the reservoir requires to spend a much long time.

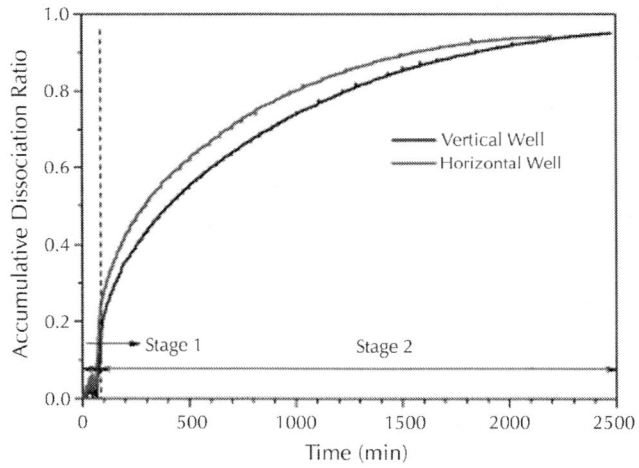

Figure 6: Evolution of accumulative dissociation ratio of gas production by depressurization.

Fig. 7 shows the profiles of the cumulative mass of water production (m_w) during the experiment. As seen, in the first stage, during the short period from 64 to 73.6 min, M_w increases from 0 to 1411 g with the horizontal well pattern. For the vertical well pattern, water production occurs from 77 min to 85 min, and M_w increases from 0 to 255 g. The cumulative mass of water production for the horizontal well is much more than that with the vertical well in the first stage. It is due to the fact that the gravity effect on the horizontal well is weaker than that on the vertical well. Therefore, the water is more easily to flow into the horizontal well and be produced out.

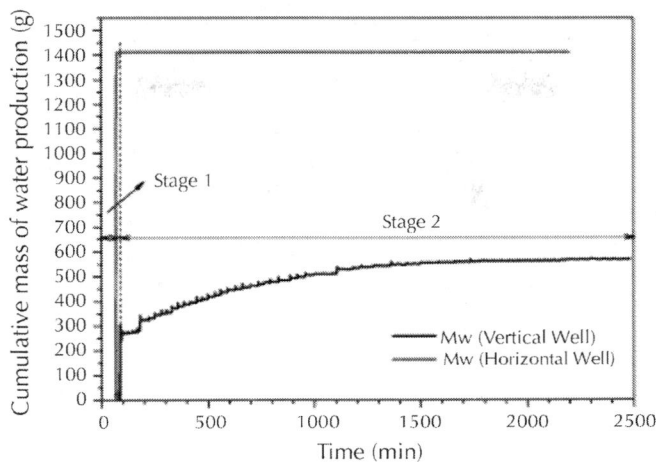

Figure 7: Evolution of cumulative mass of water production by depressurization.

In the second stage, water production rate grows over time with the vertical well pattern. It is noted that water production mainly occurs from 85 min to 1500 min with the vertical well pattern, which is because the majority of the hydrate is dissociated during this period. It is verified in Fig. 6 that the accumulative dissociation ratio increases from 20% to 85% from 85 min to 1500 min with the vertical well pattern. However, there is no water production in the whole process of the second stage with the horizontal well pattern. The reason is that the total amount of water content in the hydrate reservoir is determinate. The majority of water has been produced out in the first stage, and leaves a lot of pore space. Hence, the dissociated water in the second stage stays in the reservoir and occupies the original space of hydrate and the produced gas. It indicates the water production in the hydrate dissociation process by depressurization with the horizontal well pattern is manageable.

Thermal Stimulation

Gas production for both runs 3 and 4 are conducted by the huff and puff method. Firstly, the pressure reduces from 13.5 MPa to 6.5

MPa. Then the warm water is injected into the reservoir, and the huff and puff process starts. Because the initial depressurization stage is similar to the first stage by the depressurization method, which has been discussed in the above section. Here, we only discuss the dissociation behaviors during the huff and puff process.

The aim of this section is to evaluate the effect of the thermal stimulation on the production behaviors with the different well patterns. To eliminate the effect of depressurization on the hydrate dissociation, we set the minimum working pressure during the production stage as 6.5 MPa, which is higher than the corresponding equilibrium pressure (5.68 MPa).

Pressure and Temperature

Fig. 8a gives the evolution of the pressure and the average temperature in the whole huff and puff process.Fig. 8b shows the changes of the pressure and the average temperature in the 4th cycle, which is a typical example of the initial 10 cycles. Fig. 8c gives the changes of the pressure and the average temperature in the 21st cycle, which is a typical example from 11th to 40th cycle. As seen in Fig. 8a, Fig. 8b and Fig. 8c, there are a total of 40 huff and puff cycles for both the vertical and the horizontal well. Each single cycle includes three stages: the injection stage, the soaking stage, and the production stage. During the injection stages, the pressure in the hydrate reservoir increases on account of water and thermal injection, and the temperatures increase under the effect of heat injection and thermal diffusion in the sandy sediment. During the soaking stages, the system pressure increases in the initial ten cycles, while the increase trend of the pressures diminishes with the increase of the cycle number, and the pressures decrease from the 11th to the 40th cycle. This is because, in the first few cycles, the region influenced by the thermal stimulation increases with the injected heat diffusion from the injected well to the surroundings. It causes the increase of the hydrate dissociation rate and the increase of the amount of the produced gas, resulting in the increase of the system pressure. Meanwhile, more injected energy is lost by the absorption of the sandy sediment with the expansion of the hydrate dissociation area. In the later cycles (11th–40th cycle), the majority of the injected energy is lost by expanding from the dissociated area to the un-dissociated area, and there is little amount of hydrate is dissociated in the reservoir in the soaking stages. Therefore, the pressures decrease in the soaking stages from the 11th to the 40th cycle. During the production stages, the pressure reduces on account of gas and water production, and the temperature decreases due to the heat loss and hydrate dissociation.

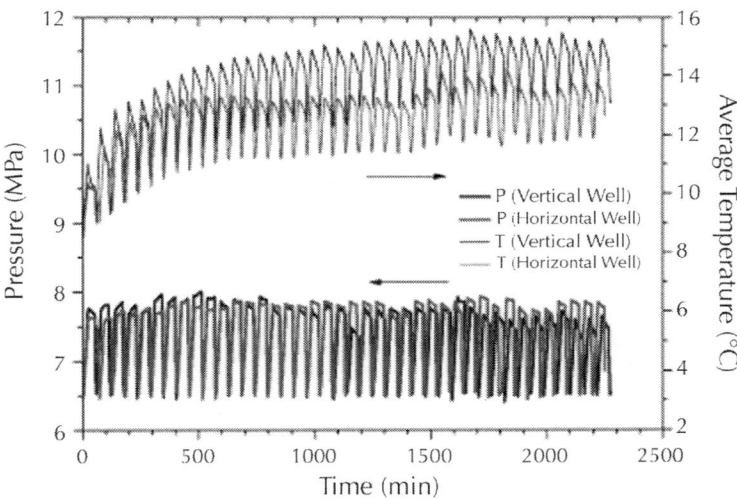

Figure 8a: Evolution of pressure and average temperature in the PHS by thermal stimulation.

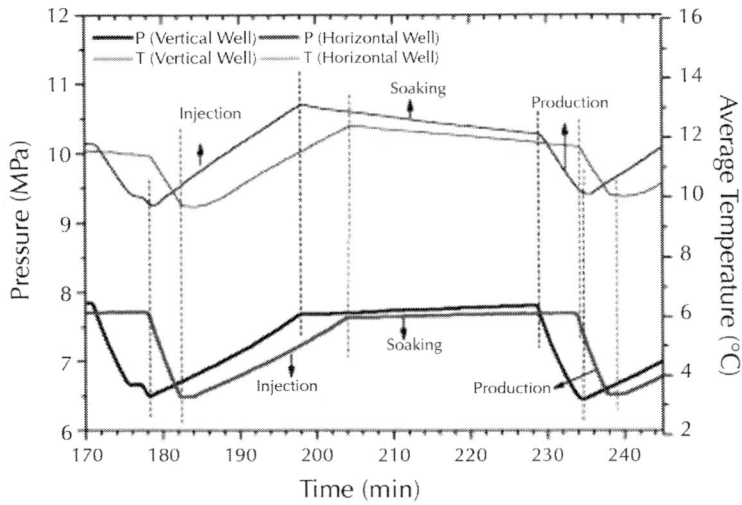

Figure 8b: Evolution of pressure and average temperature during the 4th cycle of the huff and puff process.

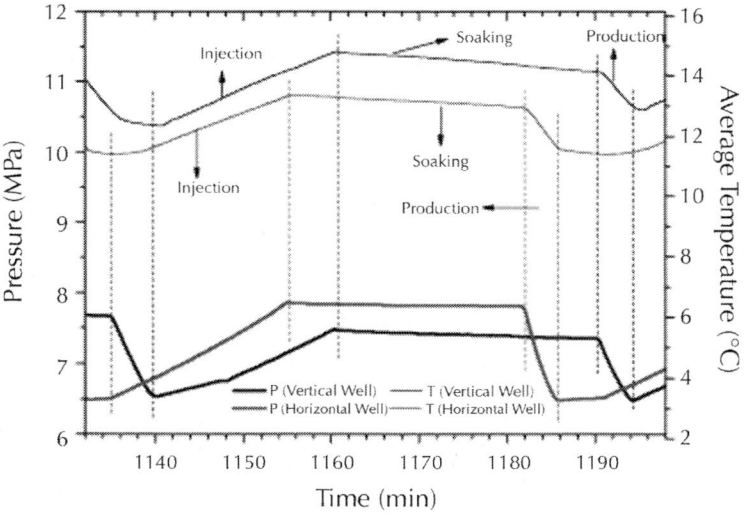

Figure 8c: Evolution of pressure and average temperature during the 21st cycle of the huff and puff process.

In addition, the amounts of the provided energy for the two runs are the same because the temperature and the injection rates of the injected water for the two runs are the same. However, it can be seen from Fig. 8athat the system temperature with the horizontal well pattern is lower than that with the vertical well pattern in the whole production process, indicating the injected heat can be more easily absorbed by the reservoir for the hydrate dissociation with the horizontal well pattern.

Fig. 9 shows the temperature distributions during the 4th cycle of the huff and puff process for both the vertical and horizontal well patterns. Fig. 9a–d and e–h represents the temperature changes for the vertical and the horizontal well pattern, respectively. Fig. 9a is the beginning of the injection stage. Fig. 9b is the endpoint of the injection stage or the initial point of the soaking stage. Fig. 9c stands for the endpoint of the soaking stage or the start of the production stage, and Fig. 9d is the termination of the production stage.Fig. 9e–h shows the temperature distribution at the corresponding stage for the horizontal well pattern. As shown in Fig. 9a–d, for the vertical well pattern, heat spreads from the center to the boundary as a cylindrical way. Fig. 9a and b depicts that the injected heat transfers from the upper part to the lower part under the effects of the gravity and the fluid flow. At the end of the soaking stage (Fig. 9c), the high-temperature part gathers at the middle-lower part in the reservoir. In the production stage, the temperature declines mainly due to the endothermic effect of hydrate dissociation and the heat absorption of the sandy sediment. The comparison of Fig. 9a and d shows that the temperature in the reservoir increases after a huff and puff cycle. However, the temperature-rising area only expands around the vertical well, and the expansion of the heat influencing region is limited. It is noted from Fig. 9b–c and f–g that, unlike the injected heat confined to the center of the reservoir with the vertical well pattern, heat transfers more uniformly with the horizontal well pattern. In addition, the heat front (heat influencing region) can reach further place with the horizontal well pattern. This is because the heat transfer rate is higher with the horizontal well pattern. In a word, the horizontal well is more favorable for heat transfer in the sandy sediment, which is in good agreement with the above discussion about the depressurization.

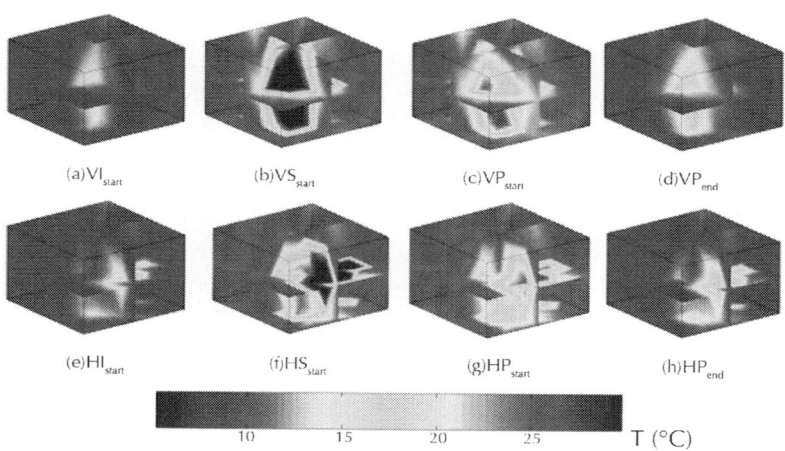

Figure 9: Distribution of the temperatures during the 4th cycle of the huff and puff process.

Gas and Water Production

Fig. 10 gives the changes of the cumulative volumes of gas production (V_p) for runs 3 and 4. As seen in Fig. 10, the amount of the produced gas with the horizontal well is larger than that with the vertical well. The reason is that the spreading rate of the area influenced by the injected heat is faster with the horizontal well pattern, and more amount of the produced gas is accumulated with the horizontal well pattern.

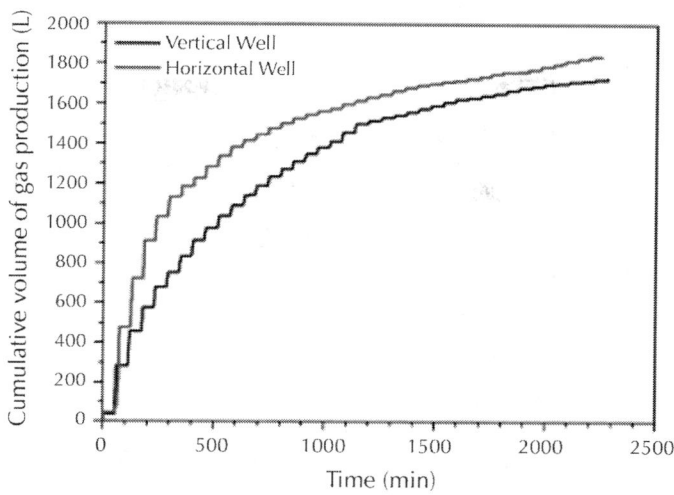

Figure 10: Evolution of cumulative volume of gas production in the PHS by thermal stimulation.

Fig. 11a, Fig. 11b and Fig. 11c give the changes of the accumulative dissociation ratio for the whole huff and puff process, the 4th and the 21st cycle, respectively. As shown in Fig. 11a, the final accumulative dissociation ratios for the vertical and horizontal well patterns are 53.35% and 55.88%, respectively. This phenomenon is in agreement with the previous study that the hydrate cannot be dissociated completely by the huff and puff method above the equilibrium pressure with single well [16]. It is obvious that the horizontal well pattern obtains a higher accumulative dissociation ratio, indicating that more hydrate is dissociated with the horizontal well pattern. The hydrate dissociation with the horizontal well pattern can be divided into three stages by the characteristics of the accumulative dissociation ratio. They are H1: the fast dissociation stage (0–5 cycle), H2: the slow dissociation stage (6–16 cycle), and H3: the little dissociation stage (17–40 cycle), respectively. The accumulative dissociation ratio during H1 increases from 0% to 33%, and the average accumulative dissociation ratio is 6.6% per cycle. Afterward, with the continuous spreading of the hydrate dissociation front, more energy is consumed by the absorption of the sandy sediment. Therefore, the hydrate dissociation rate decreases over time. Fig. 11a shows that the average accumulative dissociation ratio in H2 is 1.8% per cycle, which is only 27.2% of the average ratio in H1. During H3, the increment of the accumulative dissociation ratio is only 3.4% (from 53.4% to 57.2%), indicating that there is little hydrate dissociation in H3. This is because almost all of the injected energy is consumed by heating the sandy sediment and by the heat exchange with the boundary before the injected energy reaches the un-dissociated region of the hydrate reservoir.

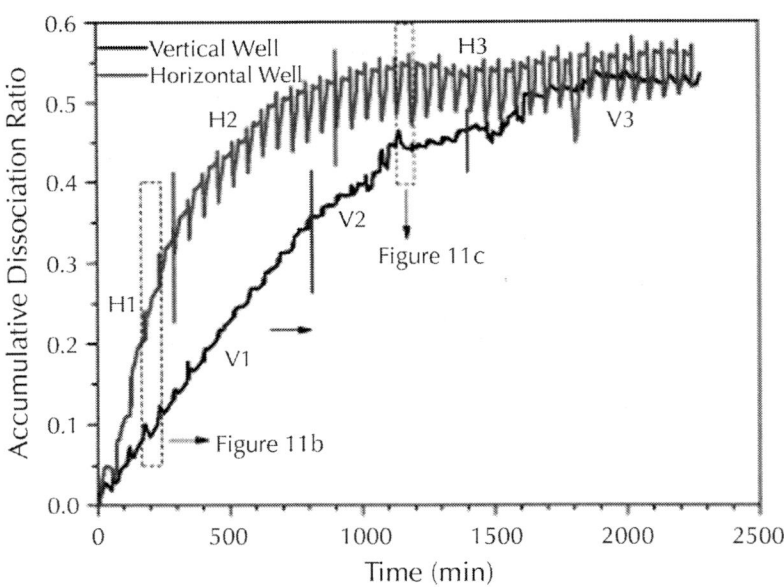

Figure 11a: Evolution of accumulative dissociation ratio of gas production in the PHS by thermal stimulation.

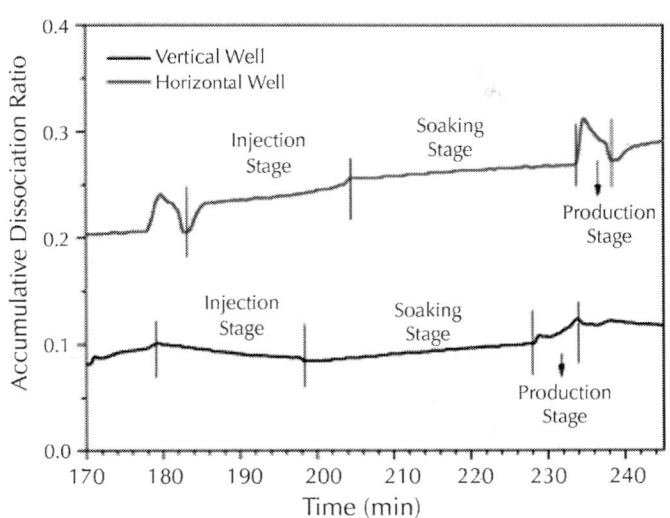

Figure 11b: Evolution of accumulative dissociation ratio for the 4th cycle.

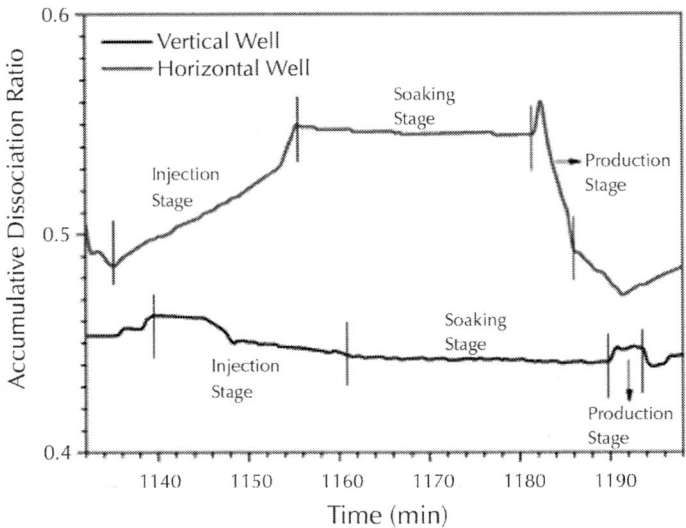

Figure 11c: Evolution of accumulative dissociation ratio for the 21st cycle.

The hydrate dissociation with the vertical well pattern can be divided into the corresponding three stages. They are named as V1 (0–14 cycle), V2 (15–34 cycle), and V3 (35–40 cycle), respectively. The average accumulative dissociation ratios for V1–3 are 2.56%, 0.86%, and 0.04% per cycle, respectively. It can be seen from Fig. 11a that the values of the accumulative dissociation ratios at the ends of the first two stages are almost the same for the two well patterns. While the duration of the first two stages is much longer with the vertical well pattern. Additionally, the average accumulative dissociation ratio per cycle with the vertical well pattern is much smaller than that in the corresponding stage with the horizontal well pattern. Thus, it indicates that the hydrate is dissociated more quickly with the horizontal well pattern.

As shown in Fig. 11b, the accumulative dissociation ratio for the vertical well decreases continuously from 9.79% to 8.69% during the injection stage of the 4th cycle, which means that the amount of hydrate increases in the reservoir. The reason is that, although the hydrate in the vicinity of the injection well is dissociated under thermal stimulation, the hydrate re-formation occurs in the un-dissociated region on account of the pressurization effect caused by the hot water injection, and the amount of hydrate formation is more than that of hydrate dissociation [29]. However, the accumulative dissociation ratio with the horizontal well during the injection stage increases from 20.56% to 25.66%, which indicates the effect of thermal stimulation (inducing hydrate dissociation) is stronger than the pressurization effect (causing hydrate formation) with the horizontal well pattern.

Actually, hydrate dissociation and formation simultaneously occur during the whole huff and puff process. The increase or decrease of the accumulative dissociation ratio is determined by the competition of the amount of hydrate dissociation and formation. In order to explain this phenomenon, Fig. 12 gives the decomposition boundary (where the temperature is above the equilibrium temperature) and the temperature distribution at the 10th minute in the 4th injection stage. Fig. 12a and c represents the vertical and the horizontal well pattern, respectively. Fig. 12b and d are the sectional drawing for the vertical and horizontal well, respectively. It is noted in Fig. 12 that the temperatures of the region outside the decomposition interface is lower than the equilibrium temperature, which satisfies the condition for hydrate formation. Meanwhile, the temperatures inside the decomposition interface are higher than the equilibrium temperature, which causes hydrate dissociation. The rate of hydrate dissociation is determined by the moving rate of the decomposition interface and the amount of the hydrate swept by the decomposition interface. Moreover, the rate of the hydrate formation is determined by the contact area of gas and water, the rate of fluid flow, and the temperature driving force for the hydrate formation.

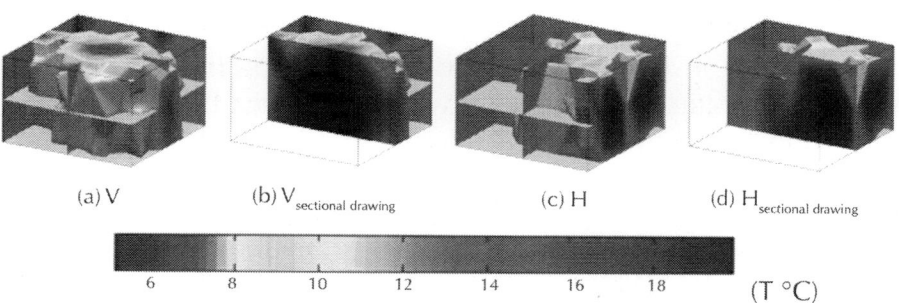

(a) V   (b) V$_{sectional\ drawing}$ (c) H   (d) H$_{sectional\ drawing}$

6   8   10   12   14   16   18   (T °C)

Figure 12: Decomposition boundaries and spatial distribution of temperature at the 10th minute of the injection stage of the 4th cycle for runs 3 and 4.

As also shown in Fig. 11b, during the soaking stage of the 4th cycle, the accumulative dissociation ratios for the two well patterns increase, which is caused by the amount of hydrate dissociation is larger than that of the hydrate formation. During the production stage of the 4th cycle, the accumulative dissociation ratio for the vertical well pattern increases from 10.89% to 12.34%. However, it is interestingly noted that for the horizontal well in this stage, the accumulative dissociation ratio firstly increases from 26.87% to 31.46% (234–235 min), and then it decreases from 31.46% to 27.17% (235–238.5 min). This is because the amount of hydrae formation is smaller than that of hydrate dissociation from 234 to 235 min with the horizontal well. Afterward, the rate of the hydrate formation is faster than that of the hydrate dissociation. This phenomenon indicates the fluid flow in the sediment is faster with the horizontal well, and the contact of gas and water is more sufficiently, which is favorable for the hydrate formation. Furthermore, Fig. 11a shows that the hydrate formation phenomenon during the production stage tends to be remarkable with the increase of the huff and puff cycle in the horizontal well. In addition, the increase of the accumulative dissociation ratio is limited (from 47.31% to 55.88%) in the later 30 cycles. Consequently, the later 30 cycles is uneconomical for hydrated dissociation and other method such as depressurization can be employed to dissociate the hydrate during the later period.

As shown in Fig. 11c, the variation trend of the accumulative dissociation ratio for the vertical well is very small during the 21st cycle, indicating that the effect of thermal stimulation on the undissociated region diminishes with the increase of the cycle number. In the 21st cycle, both the changes of the injection and production stage with the vertical and the horizontal wells are similar to that in the 4th cycle. The difference is that during the soaking stage of the 21st cycle, the dissociation ratio for the vertical well reduces from 44.44% to 44.13%, and it decreases from 55.04% to 54.53% with the horizontal well. This indicates that the amount of hydrae dissociation is smaller than that of hydrate formation for both the vertical and the horizontal wells in the reservoir during the soaking stage.

Fig. 13 shows the cumulative mass of water production (m_w) for runs 3 and 4, respectively. The water injection rates and the durations of water injection for the two runs are 0.2 L/min and 20 min, respectively. Thus, a total of 4 L water is injected into the reservoir per cycle, and the total amounts of the water injection during the 40 cycles for both the two well patterns are 160 L. As shown in Fig. 13, the difference of water production for the vertical well pattern (154,643 g) and that for the horizontal well pattern (145,999 g) is small. This is because the amount of produced water is mainly originated from water injection other than that from hydrate dissociation in this case.

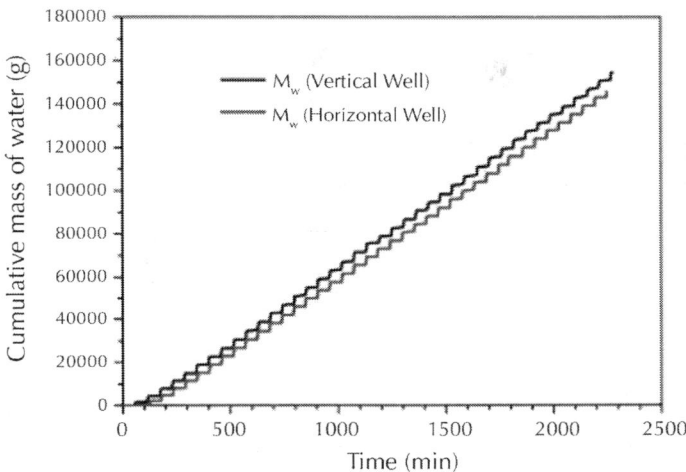

Figure 13: Evolution of cumulative volume of water production in the PHS by thermal stimulation.

Energy Ratio and Thermal Efficiency

Energy ratio and thermal efficiency are employed to evaluate the production efficiency of the hydrate reservoir in the PHS. The energy ratio is defined as the ratio of the combustion heat of the methane gas to the total input energy, which can be described as:

$$\xi = \frac{V_P M_{gas}}{C_w M_{inj}(T_0 - T)}$$

(10)

where V_p is the cumulative volume of the produced gas, and M_{gas} (37.6 MJ/m^3) is the combustion heat of the methane gas. C_w (4.2 × 10^3 J kg^{-1} k^{-1}) represents the specific heat of water. M_{inj} is the cumulative mass of the injected water. T_0 and T are the initial temperature and the ambient temperature (8 °C), respectively.

The thermal efficiency is the ratio of the heat applied for hydrate dissociation to the total injected energy, which can be described by the following equation

$$\eta = \frac{m_{diss} M_{hyd}}{C_w M_{inj}(T_0 - T)}$$

(11)

where m_{diss} is the mole of the dissociated hydrate, and the dissociation heat of the hydrate (M_{hyd}) is 54.1 kJ/mol.

Fig. 14 shows the evolution of energy ratio and thermal efficiency for runs 3 and 4. In the injection and soaking stages of the first cycle with the two well patterns, the energy ratios are 0 because there is no gas production. Subsequently, the energy ratios increase to the peak value at the end of the first cycle for the two well patterns. The maximum values of the energy ratio for the vertical and horizontal well patterns are 15.42 and 26.32, respectively. After the peak value, the energy ratios for the two patterns gradually decrease to the minimum value. The energy ratio with the horizontal well pattern is higher than the vertical well pattern in the whole process of hydrate dissociation, suggesting the horizontal well pattern is more profitable for hydrate dissociation from methane hydrate in the PHS.

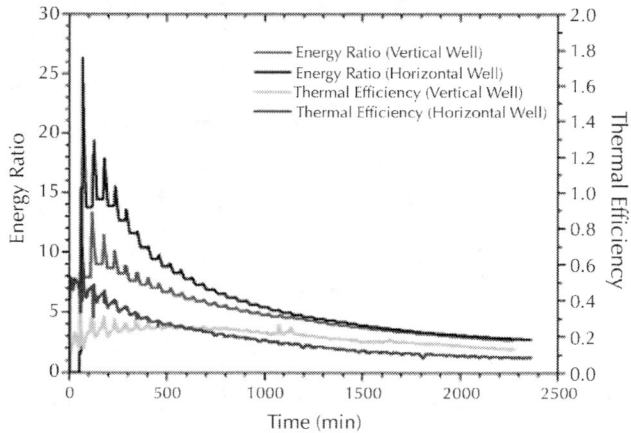

Figure 14: Evolution of energy ratio and thermal efficiency in the PHS by thermal stimulation.

In the initial 10 cycles, the thermal efficiency for run 4 (from 0.63 to 0.27) is higher than that for run 3 (from 0.32 to 0.27), which indicates that more injected energy is employed for hydrate dissociation with the horizontal well pattern. It is in good agreement with what is described in Fig. 11a, Fig. 11b and Fig. 11c that the increase rate of the accumulative dissociation ratio for run 4 is much faster than that for run 3. During the later 30 cycles (11–40 cycle), the thermal efficiency for run 4 reduces from 0.27 to 0.09 with the horizontal well, and it drops from 0.27 to 0.13 for run 3 with the vertical well. The thermal efficiencies for the two well patterns keep at quite low level, indicating that the later 30 cycles are uneconomical for hydrate dissociation. This is because most of the input energy is lost by the absorption of the sandy sediment and the exchange with the boundary. The thermal efficiency during the later 30 cycles is slightly higher with run 3. The reason is that, during the later 30 cycles, the accumulative dissociation ratio with the horizontal well only increases from 47.31% to 55.88%, and it increases from 26.94% to 53.35% with the vertical well.

Generally, the evaluations of the cumulative volume of gas production, the accumulative dissociation ratio, the energy ratio, and the thermal efficiency indicate that the horizontal well is more favorable for gas production from the hydrate reservoir in the PHS by thermal stimulation, compared with the vertical well. This advantage may be more prominent in the practical field because the horizontal well can be much longer than the vertical well under the layered distribution of the practical hydrate reservoirs.

CONCLUSIONS

In this work, the effects of vertical and horizontal well patterns on gas production performance in the PHS have been evaluated. Both the depressurization method and the thermal huff and puff method have been employed. The conclusions can be drawn as follows:

For the depressurization method, the heat transfer rate with the horizontal well is higher than that with the vertical well, the duration with the horizontal well is shorter than the vertical well, and the volume of the produced gas with the horizontal well increases faster than that with the vertical well, illustrating the horizontal well pattern has a better efficiency for hydrate dissociation. The gravity effect on the horizontal well is weaker than that on the vertical well. Therefore, the water is more easily to flow into the horizontal well and be produced out in the first stage. In the second stage, there is almost no water production with the horizontal well, because the total water content in the reservoir is determinate, and the majority of the water has been produced out in the first stage.

For the thermal stimulation method, the comparisons of the distribution of temperature show that the transfer of the heat front is only confined to the center in the PHS, and the heat transfer is more quick and uniform with the horizontal well pattern. The increase or decrease of the accumulative dissociation ratio is determined by the competition of the amount of the hydrate formation and dissociation in the process of the huff and puff. In addition, the hydrate is mainly dissociated in the initial 10 cycles, and the majority of the injected energy is lost by the absorption of the sandy sediment and the exchange with the boundary in the later 30 cycles. Moreover, the evaluations of the cumulative gas production, the accumulative dissociation ratio, the energy ratio and the thermal efficiency show that the horizontal well pattern is more favorable for hydrate dissociation than the vertical well. This advantage may be more remarkable in the practical field. It is due to the fact that the horizontal well can be much longer than the vertical well under the layered distribution of the practical hydrate reservoirs.

ACKNOWLEDGMENTS

This work is supported by National Science Fund for Distinguished Young Scholars of China (51225603), National Natural Science Foundation of China (51376183, 51406210, 51276182, and 51476174), Key Arrangement Programs of the Chinese Academy of Sciences (KGZD-EW-301-2) and Science & Technology Program of Guangzhou (2012J5100012), which are gratefully acknowledged.

REFERENCES

1. Sloan ED. Fundamental principles and applications of natural gas hydrates. Nature 2003;426(6964):353–9.

2. Klauda JB, Sandler SI. Global distribution of methane hydrate in ocean sediment. Energy Fuels 2005;19(2):459–70.

3. Moridis GJ, Collett TS, Pooladi DM, Hancock S, Santamarina C, Boswell R, et al. Challenges, uncertainties, and issues facing gas production from gas-hydrate deposits. SPE Reservoir Eval Eng 2011;14(1):76–112.

4. Makogon YF, Holditch SA, Makogon TY. Natural gas-hydrates – a potential energy source for the 21st century. J Petrol Sci Eng 2007;56(1–3):14–31.

5. Xiong LJ, Li XS, Wang Y, Xu CG. Experimental study on methane hydrate dissociation by depressurization in porous sediments. Energies 2012;5(2): 518–30.

6. Li B, Li XS, Li G, Feng JC, Wang Y. Depressurization induced gas production from hydrate deposits with low gas saturation in a pilot-scale hydrate simulator. Appl Energy 2014;129:274–86.

7. Konno Y, Masuda Y, Hariguchi Y, Kurihara M, Ouchi H. Key factors for depressurization-induced gas production from oceanic methane hydrates Energy Fuels 2010;24:1736–44.

8. Zhao JF, Zhu ZH, Song YC, et al. Analyzing the process of gas production for natural gas hydrate using depressurization. Appl Energy 2015;142:125–34.

9. Zhao JF, Liu D, Yang MJ, Song YC. Analysis of heat transfer effects on gas production from methane hydrate by depressurization. Int J Heat Mass Transfer 2014;77:529–41.

10. Li XS, Wang Y, Li G, Zhang Y, Chen ZY. Experimental investigation into methane hydrate decomposition during three-dimensional thermal huff and puff. Energy Fuels 2011;25(4):1650–8.

11. Yang X, Sun CY, Yuan Q, Ma PC, Chen GJ. Experimental study

on gas production from methane hydrate-bearing sand by hot-water cyclic injection. Energy Fuels 2010;24:5912–20.

12.   Linga P, Haligva C, Nam SC, Ripmeester JA, Englezos P. Recovery of methane from hydrate formed in a variable volume bed of silica sand particles. Energy Fuels 2009;23:5508–16.

13.   Kvamme B, Kuznetsova T. Hydrate dissociation in chemical potential gradients: theory and simulations. Fluid Phase Equilib 2004;217(2):217–26.

14.   Li S, Zhang L, Jiang X, Li X. Hot-brine injection for the dissociation of natural gas hydrates. Pet Sci Technol 2013;31(13):1320–6.

15.   Dong FH, Zang XY, Li DL, Fan SS, Liang DQ. Experimental investigation on propane hydrate dissociation by high concentration methanol and ethylene glycol solution injection. Energy Fuels 2009;23:1563–77.

16.   Jung JW, Santamarina JC. CH4–CO2 replacement in hydrate-bearing sediments: a pore-scale study. Geochem Geophys Geosyst 2010;11:1–8.

17.   Xu K, Zhao JF, Liu D, Song YC, Liu WG, Xue KH, et al. A review on experimental research on natural gas production from the hydrate with CO2. Proceedings of the asme 31st international conference on ocean, offshore and artic engineering, vol. 7; 2013. p. 1337–42.

18.   Li XS, Wang Y, Li G, Zhang Y. Experimental investigations into gas production behaviors from methane hydrate with different methods in a cubic hydrate simulator. Energy Fuels 2012;26(2):1124–34.

19.   Moridis GJ, Kim J, Reagan MT, Kim SJ. Feasibility of gas production from a gas hydrate accumulation at the UBGH2-6 site of the Ulleung basin in the Korean East Sea. J Petrol Sci Eng 2013;108:180–210.

20.   Zhao JF, Yu T, Song YC, Liu D, et al. Numerical simulation of gas production from hydrate deposits using a single vertical well by depressurization in the Qilian Mountain permafrost, Qinghai-Tibet Plateau, China. Energy 2013;52:308–19.

21. Li G, Moridis GJ, Zhang K, Li XS. The use of huff and puff method in a single horizontal well in gas production from marine gas hydrate deposits in the Shenhu Area of South China Sea. J Petrol Sci Eng 2011;77(1):49–68.

22. Su Z, Moridis GJ, Zhang K, Wu N. A huff and puff production of gas hydrate deposits in Shenhu area of South China Sea through a vertical well. J Petrol Sci Eng 2012;86–87:54–61.

23. Feng JC, Li XS, Li G, Li B, Chen ZY, Wang Y. Numerical investigation of hydrate dissociation performance in the South China Sea with different horizontal well configurations. Energies 2014;7(8):4813–34.

24. Japan completes first offshore methane hydrate production test-methane successfully produced from deepwater hydrate layers. Fire in the Ice. Methane Hydrate Newsletter 2013;13(2):1–2.

25. Phelps TJ, Peters DJ, Marshall SL, West OR, Liang LY, Blencoe JG, et al. A new experimental facility for investigating the formation and properties of gas hydrates under simulated seafloor conditions. Rev Sci Instrum 2001;72(2): 1514–21.

26. Zhou Y, Castaldi MJ, Yegulalp TM. Experimental investigation of methane gas production from methane hydrate. Ind Eng Chem Res 2009;48(6):3142–9.

27. Fitzgerald GC, Castaldi MJ, Zhou Y. Large scale reactor details and results for the formation and decomposition of methane hydrates via thermal stimulation dissociation. J Petrol Sci Eng 2012;94–95:19–27.

28. Schicks JM, Spangenberg E, Giese R, Steinhauer B, Klump J, Luzi M. New approaches for the production of hydrocarbons from hydrate bearing sediments. Energies 2011;4(12):151–72.

29. Konno Y, Jin Y, Shinjou K, Nagao J. Experimental evaluation of the gas recovery factor of methane hydrate in sandy sediment. Rsc Adv 2014(4):51666–75.

30. Li XS, Yang B, Zhang Y, Li G, Duan LP, Wang Y, et al. Experimental investigation into gas production from methane hydrate in sediment by depressurization in a novel pilot-scale

hydrate simulator. Appl Energy 2012;93:722–32.

31. Li XS, Yang B, Li G, Li B, Zhang Y, Chen ZY. Experimental study on gas production from methane hydrate in porous media by huff and puff method in Pilot-Scale Hydrate Simulator. Fuel 2012;94(1):486–94.

32. Li XS, Yang B, Duan LP, Li G, Huang NS, Zhang Y. Experimental study on gas production from methane hydrate in porous media by SAGD method. Appl Energy 2013;112:1233–40.

33. Wang Y, Li XS, Xu WY, Li QP, Zhang Y, Li G, et al. Experimental investigation into factors influencing methane hydrate formation and a novel method for hydrate formation in porous media. Energy Fuels 2013;27(7):3751–7.

34. Li XS, Zhang Y, Li G, Chen ZY, Yan KF, Li QP. Gas hydrate equilibrium dissociation conditions in porous media using two thermodynamic approaches. J Chem Thermodyn 2008;40(9):1464–74.

35. Sun SC, Ye YG, Liu CL, et al. P–T stability conditions of methane hydrate in sediment from South China Sea. J Nat Gas Chem 2011;20(5):531–6.

36. Li G, Li XS, Yang B, et al. The use of dual horizontal wells in gas production from hydrate accumulations. Appl Energy 2013;112:1303–10.

37. Feng JC, Wang Y, Li XS, Li G, Chen ZY. Production behaviors and heat transfer characteristics of methane hydrate dissociation by depressurization in conjunction with warm water stimulation with dual horizontal wells. Energy 2015;79:315–24.

38. Pang WX, Xu WY, Sun CY, Zhang CL, Chen GJ. Methane hydrate dissociation experiment in a middle-sized quiescent reactor using thermal method. Fuel 2009;88(3):497–503.

Copper Recovery from Barren Cyanide Solution by Using Electrocoagulation Iron Process

José R. Parga[1], Guillermo Tiburcio Munive[2],
Jesús L. Valenzuela[2], Víctor V. Vazquez[2], and
Gregorio González Zamarripa[3]

[1]Institute Technology of Saltillo, Saltillo, Mexico

[2]Departament of Chemistry and Metallurgy, University of Sonora, Hermosillo, Mexico

[3]Faculty of Metallurgy, University of Coahuila, Monclova Coah, México

ABSTRACT

This paper is a brief overview of the role of inducing the nucleated electro winning of copper by using iron electrodes in electrocoagulation (EC) process. Cyanide compounds are widely used in gold ore processing plants in order to facilitate the extraction and subsequent concentration of the precious metal. Owing to cyanide solution employed in gold processing, effluents generated have high contents of free cyanide as well as copper cyanide complexes, which lend them a high degree of toxicity. In this regard, two options for the treatment of cyanide barren solutions has been used; in two ways; first for cyanide destruction by oxidation with the use of the EC process, in theory, has the advantage of decomposing cyanide at the anode and collecting copper simultaneously by a sludge of copper magnetic iron. In both cases excellent performance can be achieved using the high capacity of the bipolar iron EC technology. We found that it is possible to reduce the copper cyanide complex from 720 mg·l^{-1} to below 10 mg·l^{-1} within 20 minutes.

INTRODUCTION

Due to the dwindling resources of simple cyanide extractable gold deposits, a large proportion of the gold processed in the 21st century will be recovered from complex gold ores, many of which will contain soluble copper minerals. It has been estimated that about 20% of all gold deposits have significant copper mineralization commonly associated with chalcopyrite, tetrahedrite, tennantite, enargite as well as bornite and chalcocite in certain ores [1]. It has also been found that the majority of copper minerals including copper oxides, carbonates, sulfides (with the exception of chalcopyrite) and native copper are highly soluble in cyanide solutions [2]. These copper containing minerals are problematic because, when ores containing such minerals are leached with cyanide to recover the gold, copper also dissolves to form stable copper cyanide complexes. The dissolution of copper consumes a substantial quantity of cyanide and thus, if not recovered imposes a significant financial cost on the gold mine. The presence of copper also causes other problems such as competition with gold to adsorb on carbon unless a sufficient free cyanide concentration is maintained, depletion of gold electrowinning cell efficiency, and gold losses by cementation onto certain copper minerals. Ores containing greater than 0.5% reactive copper may be generally considered uneconomical to process via conventional cyanidation due to the high reagent cost. Therefore, it is necessary to reduce the amount of copper, especially in leaching circuits, in order to increases the dissolution of gold and silver, to achieve this goal it is possible to remove the copper from barren solution after Merril-Crowe process using the electrocoagulation process (EC). This process can be interesting for the copper cyanide removal from processed solutions after the Merrill Crowe process, also has low cost of operation and investment.

COPPER CYANIDE CHEMISTRY

The major challenges to the processing of gold-copper ores using cyanidation is that of the high cyanide consumptions that are typically experienced, along with effective control of the leach, particularly when there is variable cyanide-soluble copper in the ore. It is widely accepted that gold dissolution in cyanide solutions occurs as sequence of two reactions shown in Equations (1) and (2), Elsner's equation shows that oxygen is critical for the dissolution of gold.

$$2Au + 4NaCN + O_2 + 2H_2O \rightarrow$$
$$2Na\left[Au(CN)_2\right] + 2NaOH + H_2O_2 \tag{1}$$

$$2Au + 4NaCN + H_2O_2 \rightarrow 2Na\left[Au(CN)_2\right] + 2NaOH \tag{2}$$

The stoichiometry of the process shows that 4 moles of cyanide are needed for each mole of oxygen present in solution. At room temperature and standard atmospheric pressure, approximately 8.2 mg of oxygen are present in one liter of water. This corresponds to 0.27×10^{-3} mol/L accordingly, the sodium cyanide concentration (molecular weight of NaCN = 49) should be equal to $4 \times 0.27 \times 10^{-3} \times 49 = 0.05$ g/L or approximately 0.01%. This was confirmed in practice at room temperature by a very dilute solution of NaCN of 0.01% - 0.5% for ores, and for concentrates rich in gold and silver of 0.5% - 5% [3]. Also, lime is added to keep the system at an alkaline pH of 10.5 - 11.0. Other factors affecting gold leaching kinetics are grain size, agitation speed, temperature, pressure, foreign ions and cyanicides.

Free cyanide exists as the uncomplexed cyanide ion, CN^-, and molecular hydrogen cyanide, HCN. These species are related by the acid dissociation of HCN:

$$HCN_{(aq)} = CN^- + H^+$$

(3)

The concentration of free cyanide is the sum of the CN^- and HCN concentrations, and the equilibrium diagram shown in Figure 1 illustrates the distribution.

This figure shows the proportions of free cyanide as CN^-, and HCN as a function of pH at 25°C. At pH values below 7, cyanide is predominantly present as the unionized HCN molecule, which is easily volatilized because of its high vapor pressure. The equilibrium is displaced in favor of cyanide ion formation at pH values above 7.

Hydrogen cyanide (HCN), also known as hydrocyanic acid, is a colorless gas or liquid with a boiling point of 25.7°C, a vapor pressure of 100 kPa at 26°C and Henry's Law constant of 6.4 atm/mole [4], this makes HCN very volatile. Thus, low pH, high temperature, low pressure, and intimate contact with air, all tend to increase the rate of dissipation of cyanide from solution as hydrogen cyanide. In addition to free cyanide, other complexes such as the metal cyanide complexes formed with gold, silver, copper, nickel, iron and cobalt must be considered.

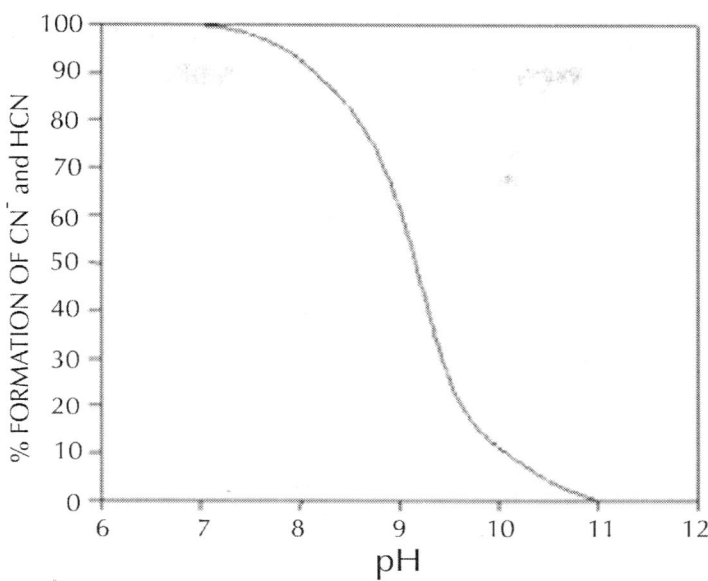

Figure 1: Equilibrium distribution diagram for cyanide as a function of pH.

In cyanidation plants all around the world, the concentration of cyanide used to dissolve gold in ores is typically higher than the stoichiometric ratio, due to the solubility of other minerals. Free cyanide produces complexes with several metallic species, especially transition metals, which show a broad variation in both stability and solubility. Many common copper minerals are soluble in the dilute cyanide solution under typical of leach conditions found in the gold cyanidation process. Minerals such as azurite and malacite, are rapidly leached and are soluble in dilute cyanide solutions.

Enargite and chalcopyrite leach more slowly but are sufficiently soluble to cause excessive cyanide loss and contamination of the pregnant leach solutions. In reactions in aqueous solutions the cupric ion is rapidly converted to cuprous form and then copper forms a series of extremely stable soluble complexes in cyanide such as:

$$Cu^+ + CN^- = CuCN \tag{4}$$

$$CuCN + CN^- = Cu(CN)_2^- \tag{5}$$

$$Cu(CN)_2^- + CN^- = Cu(CN)_3^{2-} \tag{6}$$

$$Cu(CN)_3^{2-} + CN^- = Cu(CN)_4^{3-} \tag{7}$$

Under typical gold cyanidation conditions $Cu(CN)^{2-}_3$ has been shown to be the dominant species from the EhpH diagram for the copper-cyanide-water system [4,5]. The high consumption of cyanide during the cyanidation of copper-gold ores is due to the fact that copper forms complexes of high coordination numbers with cyanide (Reaction 3 to 6), $Cu(CN)^{2-}_3$ in particular. Therefore, hydrometallurgical treatment of these ores by cyaniding as a rule gives rise to a series of difficulties associated with increase in the cyanide consumption and decrease in the dissolution rate of gold and silver, and in the cementation process. This precipitate is of low quality, because the copper is precipitated along with gold and silver, resulting in a higher consumption of zinc dust, fluxes in the smelting of the precipitate and shorter life for crucibles.

In this regard a study is proposed to remove copper cyanide ions with, a very promising electrochemical treatment technique, which does not require chemical additions. This process is electrocoagulation (EC). The EC process operates on the principle that coagulation of copper cyanide ions from barren solutions from the MerrillCrowe process is caused by the combined effects of electrolysis gases (H_2 and O_2) and the electrolytic production of cations from the iron anodes that corrode during electrolysis.

ELECTROCOAGULATION FUNDAMENTALS

The EC process operates on the principle that the cations produced electrolytically from iron and/or aluminum anodes enhance the coagulation of contaminants from an aqueous medium. Electrophoretic motion tends to concentrate negatively charged particles in the region of the anode and positively charged ions in the region of the cathode. The consumable, or sacrificial, metal anodes are used to continuously produce polyvalent metal cations in the vicinity of the anode. These cations neutralize the negative charge of the particles carried toward the anodes by electrophoretic motion, thereby facilitating coagulation. In the flowing EC techniques, the production of polyvalent cations from the oxidation of the sacrificial anodes (Fe and Al) and the electrolysis gases (H_2 and O_2) works in combination to flocculate the coagulant materials [6], the gas bubbles produced by the electrolysis carry the pollutant to the top of the solution where it is concentrated, collected and removed. Figure 2 illustrates the schematic diagram of the process.

Generally, in the EC process bipolar electrodes are used. Pretorius [7] and Mameri [8] have reported on the use of cells with bipolar electrode arranged in series. These cells, operated at relatively low current densities, produce iron or aluminum coagulant in a more effective, fast and economical manner when compared to chemical coagulation. The savings result from a reduction in the time necessary for the treatment and an increase in the anodic area improved the efficiency of the electrolysis [7, 8]. An electrocoagulation cell with bipolar electrodes connected in series is shown in Figure 3. The inner electrodes are bipolar, that is they carry both positive and negative charges on opposing faces. These charges develop on the electrode surfaces and are opposite in sign of the charge carried by the parallel electrode surface (refer to Figure 3). During electrolysis the positive side of these bipolar electrodes undergoes anodic reactions, while on the negative side, cathodic reactions occur.

The released ions neutralize the charge of the particles and thereby initiate coagulation. The bipolar arrangement reduces the time needed for the treatment due to the increase in surface area mentioned above. This arrangement also has the practical advantage of simplified set-up in that only two monopolar electrodes are connected to the electric power source with no interconnections between the inner bipolar electrodes.

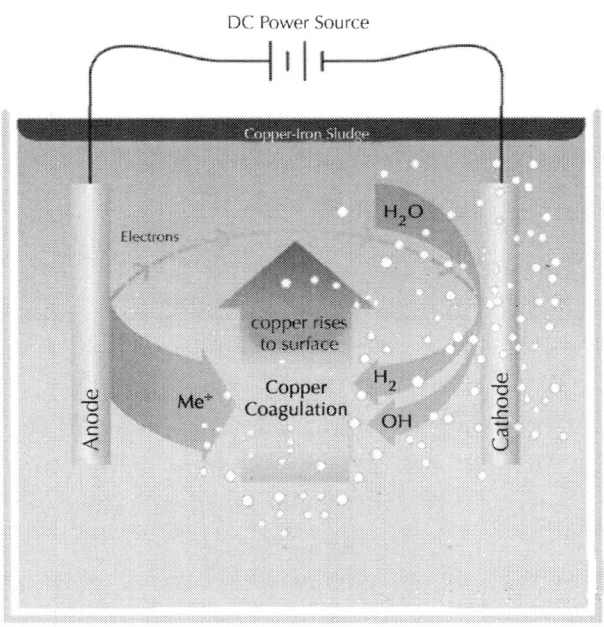

Figure 2: Schematic diagram of the EC process.

The Chemical Reactions of the Electrocoagulation Process

The chemicals reactions that have been proposed to describe the mechanism of EC for the production of $H_{2(g)}$ and $OH^-_{(ac)}$ (cathode) and $H^+_{(ac)}$ (anode) [9] are:

$$Fe \rightarrow Fe^{3+} + 3e^- \tag{8}$$

$$Fe(OH)_2^+ + H_2O \leftrightarrow Fe(OH)_3 + H^+ \tag{9}$$

$$H^+ + 2e^- \leftrightarrow H_{2(g)} \uparrow \tag{10}$$

$$Fe(OH)_2^+ + e^- \leftrightarrow Fe(OH)_{2(aq)} \tag{11}$$

$$Fe(OH)_{2(aq)} + H_2O \leftrightarrow Fe(OH)_3^- + H^+ \tag{12}$$

$$Fe(OH)_3^- \leftrightarrow Fe(OH)_{3(aq)} + e^- \tag{13}$$

Overall reaction.

$$6Fe + (12 + x)H_2O \leftrightarrow \frac{1}{2}(12 - x)H_{2(g)} \uparrow$$
$$+ xFe(OH)_3 \cdot (6 - X)Fe(OH)_{2(s)} \tag{14}$$

The pH of the medium usually rises as a result of this electrochemical process and the Green Rust formed [xFe(OH)$_3$ * (6 – X)Fe(OH)$_{2(s)}$] remains in the aqueous stream as a gelatinous suspension, which can remove the gold and silver from pregnant cyanide rich solutions, either by complexation or by electrostatic attraction followed by coagulation and flotation.

Formation of rust (dehydrated hydroxides) occurs a while after the process, as shown in the following:

$$2Fe(OH)_3 \leftrightarrow Fe_2O_3 + 3H_2O \quad \text{(hematite)} \tag{15}$$

$$Fe(OH)_2 \leftrightarrow FeO + H_2O \tag{16}$$

$$2Fe(OH)_3 + Fe(OH)_2 \leftrightarrow Fe_3O_4 + 4H_2O \quad \text{(magnetite)} \tag{17}$$

$$Fe(OH)_3 \leftrightarrow FeO(OH) + H_2O \quad \text{(goethite, lepidocrocite)} \tag{18}$$

A schematic representation of these reactions in an EC process, using iron electrodes, is shown in Figure 4. As mentioned above, the gas bubbles produced by electrolysis carry the copper along with the sludge to the top of the solution where it is collected and removed [10]. However, it is the reactions of the metal ions that enhance the formation of the coagulant. The metal cations are hydrolyzed, releasing hydrogen ions that result in hydrogen evolution at the cathode, to yield both soluble and insoluble hydroxides that will react with or adsorb copper from the cyanide solution and also contribute to coagulation by neutralizing the negatively charged colloidal particles that may be present at neutral or alkaline pH.

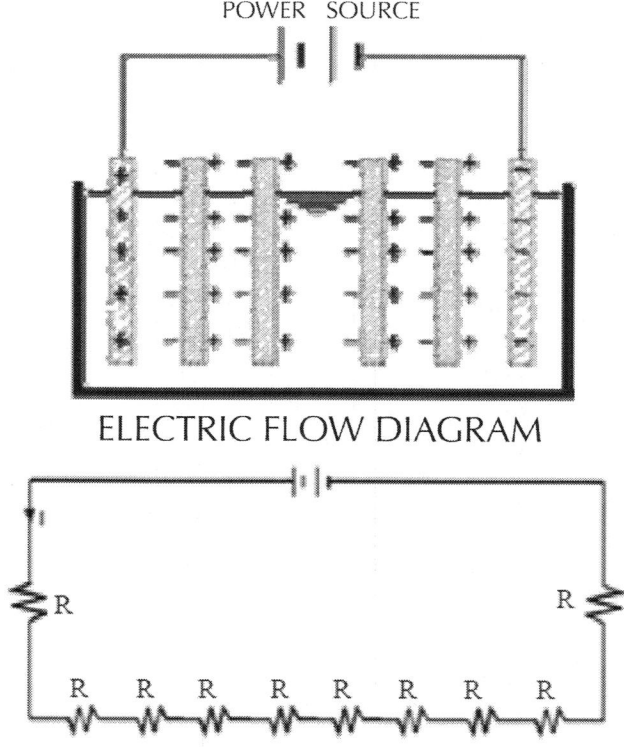

POWER SOURCE

ELECTRIC FLOW DIAGRAM

Figure 3: A schematic representation of a bipolar electrocoagulation cell and the equivalent electrical circuit (resistors reflects electrode surfaces).

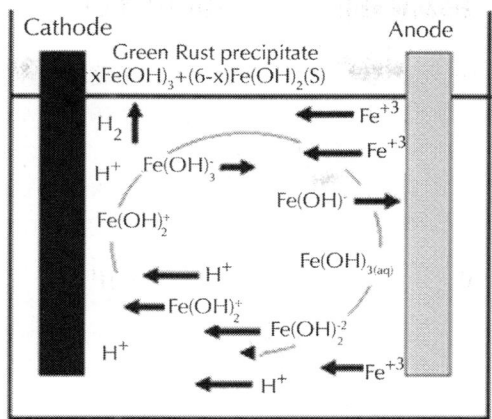

Figure 4: An illustration of the EC mechanism (arrow indicate the migration of ions, the H_2 evolution and the formation of green rust).

This enables the particles to approach closely and agglomerate under the influence of Van der Waals attracttive forces. The pH of the medium rises as a result of this electrochemical process and the $Fe(OH)_{n(s)}$ formed remains in the aqueous stream as gelatinous suspension, which can remove the $Cu(CN)^{2-}_3$ from the barren solution, either by complexation or by electrostatic attraction followed of coagulation and flotation [11]. Generally, in the EC process, bipolar electrodes are used. It has been reported that cells with bipolar electrodes, connected in series and operating at relatively low current densities [12], produce iron or aluminum coagulant more effecttively, more rapidly and more economically when compared to chemical coagulation.

EXPERIMENTAL DETAILS

The experimental work was performed using a barren solution from the Merrill Crowe process containing an average in the range of 660 - 712 ppm of copper. EC experiments were performed using a 600 ml Pyrex beaker glass (Figure 5), equipped with two iron electrodes (10 cm × 2.5 cm) with an electrode separation of 5 mm. A regulated power supply (model Steren PRL-25) to provide the necessary energy to the electrocoagulation cell, magnetic stirrer (Model PC-310, Corning) with 100 rpm was used, and the initial and final pH was taken with a pH meter (VWR Scientific 8005), filtration was performed with Whatman filter paper No.42. To determine the adsorption copper in iron hydroxide species generated at the anode, a barren solution after the cementation process was used that was provided by the William Mine Co. This solution contained an average of 0.02 mg/L gold and 0.1 mg/L silver along with amounts of zinc, lead. Among others, the analysis was performed according to EPA 200.7 (EPA 200.7 is an analytical method for detection of metals and trace elements by ICP/atomic emission spectrometry). The solution and the solids were separated by filtration through filter paper, the sludge from the EC was dried for 8 hrs in an oven at 80°C. The tables below show the initial conditions of the EC test to remove copper.

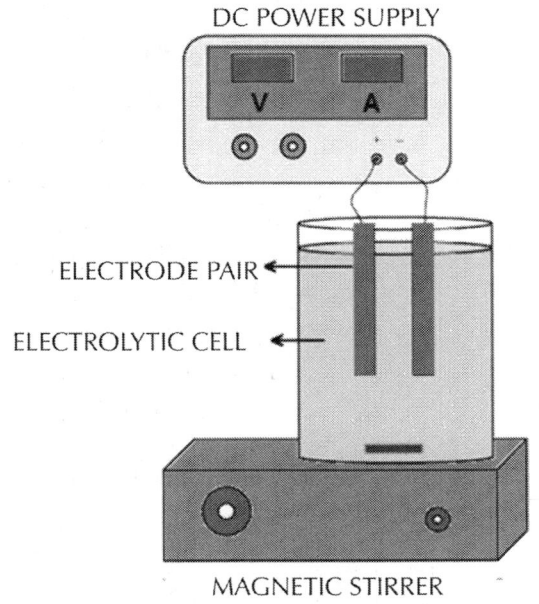

Figure 5: EC schematic diagram of the experimental setup.

Optimization of Parameters

In order to find the optimum parameters of the EC process for the removal of copper, experiments were carried out by changing the pH of the solution, residence time in the EC cell and voltage and amperage.

RESULTS

Tables 1-3 show the results for tests performed with 4 grams/liter of sodium chloride.

From these results it was determined that the optimum parameters were: pH 8, residence time of 20 minutes and 4 grams/liter of NaCl, this achieved 99% copper removal. Also, when the time increased from 15 to 20 minutes the removal of copper increased from 92% to 99%, this occurs in the pH range from 8 to 9 approximately, this coincides with the production of the magnetic iron, Fe_3O_4, which has magnetic properties that accelerates the process of adsorption of metals, the adsorption rate is then physically, because it is caused by the magnetic forces of the magnetite into copper, without altering their chemical composition. This removal of copper also can explain with a decrease in the zeta potential on iron hydroxides which causes a decrease in repulsive forces between the particles, generated collision between particles thus favors the formation of flocs which float to the water surface through micro bubbles generated from oxygen and hydrogen from the iron electrodes. Also, the advantages of the EC process is the decomposing of cyanide at the anode, where the anodic oxidation of cyanide is proportional to the alkalinity of the electrolyte and consistent with the following mechanism:

$$CN^- + 2OH \rightarrow CNO^- + H_2O + 2e^- \qquad (18)$$

$$2CNO^- + 4OH^- \rightarrow 2CO_2 + N_2 + 2H_2O + 6e^- \qquad (19)$$

$$CNO^- + 2H_2O \rightarrow NH_4 + CO_3^- \qquad (20)$$

Figure 6 shows the graph of timeelimination where it is observed the behavior of the removal of copper.

Table 1: Results obtained for pH 8 and 4 g/liter of NaCl

Initial Conditions of EC Test				Sludge Chemical Essay				
Time (min)	Voltage (volt)	Current (A)	%Cu	%Fe	$[Cu]_{initial}$(ppm)	$[Cu]_{final}$(ppm)	%Elemination
5	12.1	10.3	4.6	21.3	660	328	50
10	12.3	10.5	6.2	23	664	188	72
15	11.5	10.9	9.4	24	712	70	90
20	12.4	10.1	13	27	716	5	99

Table 2: Results obtained for pH 9 and 4 g/liter of NaCl

Initial Conditions of EC Test				Sludge Chemical Essay				
Time (min)	Voltage (volt)	Current (A)	%Cu	%Fe	$[Cu]_{initial}$(ppm)	$[Cu]_{final}$(ppm)	%Elemination
5	12.1	10.3	4.4	22	673	428	36
10	12.3	10.5	7	23.5	677	190	76
15	11.5	10.9	10	24	679	80	88
20	12.4	10.1	12.8	27.5	701	10	98

Table 3: Results obtained for pH 10 and 4 g/liter of NaCl

Initial Conditions of EC Test				Sludge Chemical Essay				
Time (min)	Voltage (volt)	Current (A)	%Cu	%Fe	$[Cu]_{initial}$(ppm)	$[Cu]_{final}$(ppm)	%Elemination
5	12.1	10.3	4.8	21.5	667	424	36
10	12.3	10.5	6.8	22	686	186	73
15	11.5	10.9	10.8	24	680	92	86
20	12.4	10.1	13	28	705	15	98

PRODUCT CHARACTERIZATION

In order to identify the iron species present, Scanning Electron Microscope (SEM/EDX) was used to characterize the solid products formed during the EC process for removal of copper with iron electrodes.

Scanning Electron Microscopy (SEM/EDAX)

Figure 7 shows SEM image and Figure 8 shows EDAX of copper adsorbed on iron species. These SEM and EDAX results show that the surfaces of these iron oxide/ oxyhydroxide particles were coated with a layer of copper.

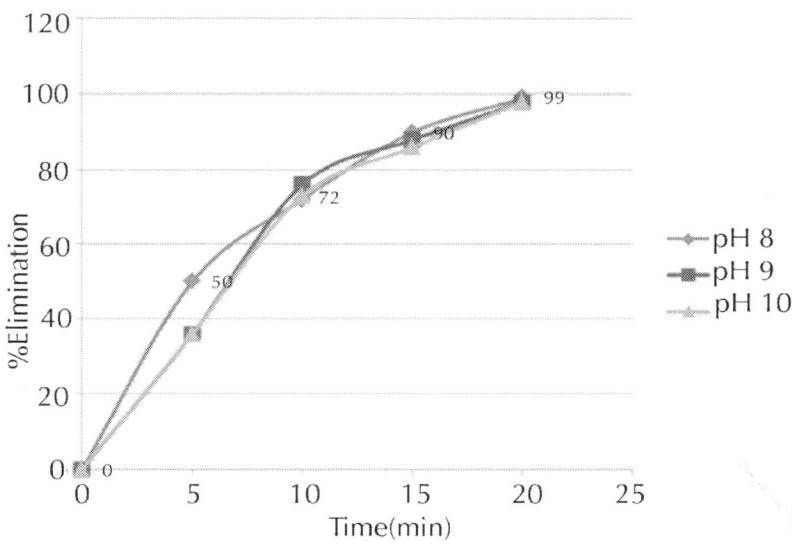

Figure 6: % Elimination of copper at pH 8, 9 and 10.

Figure 7: SEM image.

X-Ray Diffraction

Diffractograms were obtained with a Bruker AXS D4 Endeavor diffractometer operating with a Cu-K radiation source filtered with a graphite monochromator (λ = 1.5406 Å). The samples were wet ground to a fine powder (isopropyl alcohol from Sigma-Aldrich) and pressed into a sample holder. The XRD scans were recorded from 20° to 80° 2θ, with 0.02° step-width and with a 10 s counting time for every step-width (increment). Experiments were run at 40 kV and 40 mA power. Figure 9, shows a diffractogram of the filtered solid products (the feed solution contained 700 ppm of copper and the pH of the solution after EC was ~10).

This figure indicates the presence of magnetite, geothite, iron hydroxide oxide, and lepidocrocite in the solid products.

CONCLUSIONS

Electrocoagulation process was carried out to obtain a 99% of copper removal. The optimal operations conditions were: pH

= 8, residence time: 20 minutes and 4 g/l of sodium chloride as conductivity modifier. The solid product obtained from EC process was 13% copper and 24% iron. The EC process does not generate any smell in the process of elimination of copper from the barren solution. During the EC process is not necessary to add any reagent (except Nacl), since the coagulating agent is generated in situ.

We found that it is possible to reduce the copper cyanide complex from 720 mg·l⁻¹ to below 10 mg·l⁻¹ within 20 minutes.

Also, the results of this study suggest that EC produces magnetic particles of magnetite. lepidocrocita and amorphous iron oxyhydroxide species that can be used to removal copper. The Scanning Electronic Microscopy, techniques demonstrate that the formed species are of magnetic type, like lepidocrocite and magnetite which adsorbed the copper particles on his surface due to the electrostatic attraction between both metals.

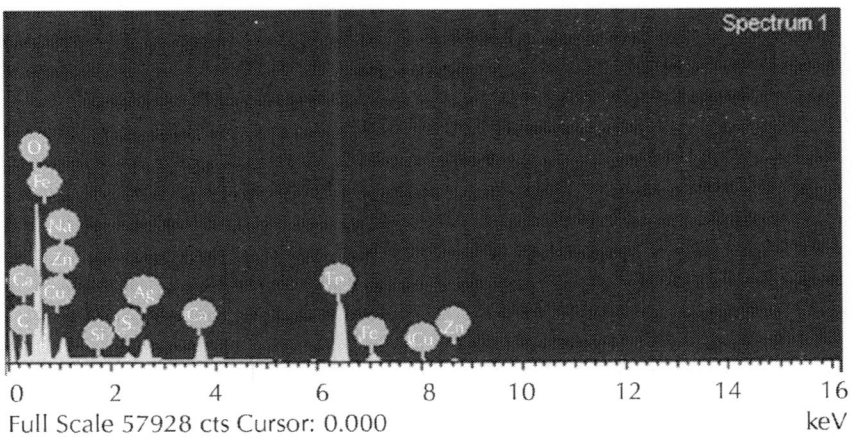

Figure 8: EDAX Analysis.

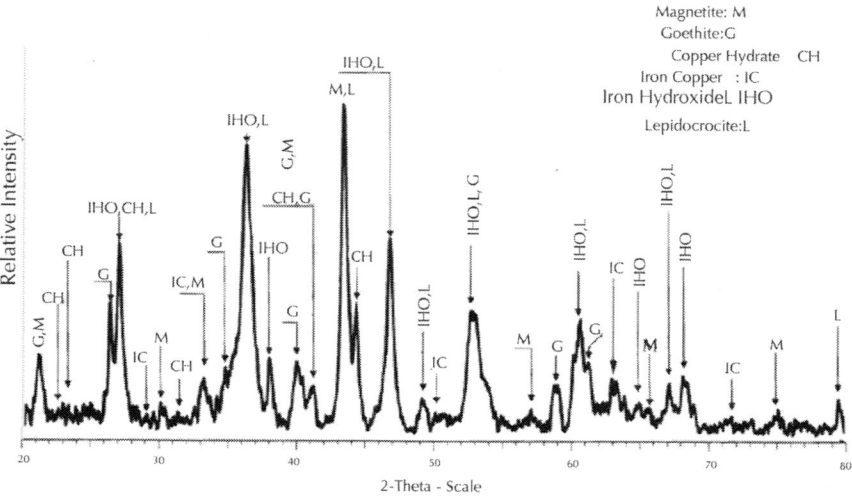

Figure 9: X-ray diffraction spectrum obtained from the EC product at pH = 10.

ACKNOWLEDGMENTS

The authors wish to acknowledge support of this project to the National Council of Science and Technology (CONACYT) and to Dirección General de Educación Superior Tecnológica (DGEST).

REFERENCES

1.   D. M. Muir, S. R. LaBrooy and C. Cao, "Recovery of Gold Fromcopper-Bearing Ores," In: R. J. Harden, Ed., Gold Forum Ontechnology and Practic, World Gold, 1989.

2.   J. O. Marsden abd C. I. House, "The Chemistry of Gold Extraction," 2nd Edition, Society for Mining, Metallurgy and Exploration, Inc., Littleton, 2006.

3.   J. R. Parga, S. S. Shukla and D. L. Cocke, "Photocatalytic Detoxification of Cyanide and Recovery of Titanium Dioxide

by Electrocoagulation," Research Journal of Chemistry and Environment, Vol. 9, No. 1, 2005, pp. 60-63.

4.   S. T. Mudder, "The Chemistry and Treatment of Cyanidation Wastes," Mining Journal Books Limited, London, 1991, pp. 277-278.

5.   M. D. Adams and R. Lawrance, "Biogenic Sulphide for Cyanide Recycle and Copper Recovery in Gold-Copper Ore Processing," International Workshop on Process Hydrometallurgy, Hydroprocess Brisbone, 2008, pp. 14-16.

6.   O. Asare, K. Xue and T. Ciminelli, "Solution Chemistry of Cyanide Leaching Systems. In Precious Metals Mineralogy," Extraction and Processing-Proceedings of and International Symposium, 1984, pp. 173-197.

7.   N. Mameri, A. R. Yeddou, H. Lounici, D. Belhocine, H. Grib and B. Bariou, "Defluoridation of Septentrional Sahara Water of North Africa by Electrocoagulation Process Using Bipolar Aluminum Electrodes," Water Research, Vol. 32, No. 5, 1998, pp. 1604-1612.doi:10.1016/S0043-1354(97)00357-6

8.   W. A. Pretorius, W. G. Johannes and G. G. Lempert, "Electrolytic Iron Flocculant Production with a Bipolar Electrode in Series Arrangement," Water South Africa, Vol. 17, No. 2, 1991, pp. 133-138.

9.   G. Pavas, M. P. Camargo, C. Jones and V. T. Pineda, "Oxidación Fotocatalítica de Cianuro," Universidad EAFIT, Medellín, 2005, pp. 56-58.

10.   J. R. Parga, H. M. Casillas, V. Vazquez and J. L. Valenzuela, "Cyanide Detoxification of Mining Wastewaters with TiO_2 Nanoparticles and Its Recovery by Electrocoagulation," Chemical Engineering and Technology, Vol. 32, No. 12, 2009, pp. 1901-1908.doi:10.1002/ceat.200900177

11.   G. Vicuña and I. Tuñon, "Apuntes de Química Avanzada," Departamento de Química-Física. Universidad de Valencia, 2006, pp. 103-115.

12.   A. G. Gupta and S. Kundu, "Adsorptive Removal of As(III) from Aqueous Solution Using Iron Oxide Coated Cement (IOCC):

Evaluation of Kinetic Equilibrium and Thermodynamic Models," Separation and Purification Technology, Vol. 51, No. 2, 2006, pp. 165-172.doi:10.1016/j.seppur.2006.01.007

A Comparative Study of Flowback Rate and Pressure Transient Behavior in Multifractured Horizontal Wells Completed in Tight Gas and Oil Reservoirs

Majid Ali Abbasi[a], Daniel Obinna Ezulike[a], Hassan Dehghanpour[a], and Robert V. Hawkes[b]

[a]School of Mining and Petroleum Engineering, Department of Civil and Environmental Engineering, Markin/CNRL Natural Resources Engineering Facility, University of Alberta, Edmonton, AB T6G 2W2, Canada

[b]SPE, Trican Well Services, Canada

ABSTRACT

Tight reservoirs stimulated by multistage hydraulic fracturing are commonly characterized by analyzing the hydrocarbon production data. However, analyzing the available hydrocarbon production data can best be applied to estimate the effective fracture–matrix interface, and is not enough for a full characterization of the induced hydraulic fractures. Before putting the well on flowback, the induced fractures are filled with the compressed fracturing fluid. Therefore, analyzing the early-time rate and pressure of fracturing water and gas/oil should in principle be able to partly characterize the induced fractures, and complement the conventional production data analysis.

We construct basic diagnostic plots by using two-phase flowback data of three multifractured horizontal wells to understand the physics of flowback. Analysis of flow rate plots suggests three separate flowback regions based on the relative values of water and gas/oil flow rate. In the first region, water production dominates while in the third region gas/oil production dominates. In the second region, water production drops and gas/oil production ramps up. The cumulative water production (CWP) plots show two distinct water recovery periods. Before gas/oil breakthrough, CWP linearly increases with time. After breakthrough, CWP increases with a slower rate, and reaches to a plateau for the oil well. We also develop a simple analytical model to compare the pressure/rate transient behavior of the three flowback cases. This work demonstrates that rate and pressure, carefully measured during the flowback operations, can be interpreted to evaluate the fracturing operations and to complement the conventional production data analysis for a more comprehensive fracture characterization.

INTRODUCTION

The amount of hydrocarbon stored in previously inaccessible shale and tight reservoirs is significantly higher than that stored in conventional reservoirs (Zahid et al., 2007 and Abdelaziz et al., 2011). Recent advances in horizontal drilling and multistage hydraulic fracturing have unlocked these challenging hydrocarbon plays. Characterizing the induced fracture network is important for evaluating the fracturing operation, and predicting the reservoir performance. Various mathematical models have been proposed for analyzing the hydrocarbon production data for the purpose of characterizing the fractured horizontal wells. The fracture–matrix interface and fracture half-length are usually determined by analyzing the hydrocarbon production data. The dual porosity model has been extended for analyzing the fractured horizontal wells (Bello, 2009, El-Banbi, 1998, Medeiros et al., 2008, Medeiros et al., 2010 and Ozkan et al., 2010). The available hydrocarbon production data mainly match the late linear transient part of the type curves, which relates to the fluid transfer from the matrix into the fracture. This match can be interpreted to determine the effective fracture half-length. However, a full characterization of the fracture network by only analyzing the hydrocarbon data is challenging because:

- The early-time oil or gas production data is usually unavailable or of low quality for history matching.
- The induced fracture network is initially filled with compressed fracturing fluid not hydrocarbon. Therefore, analyzing the hydrocarbon data for determining the fracture storage capacity can be misleading.
- Production data analysis does not account for the fractures, which are filled with water and do not contribute to the hydrocarbon flow.

Conventional rate transient methods have been applied for analyzing the flowback data. For example, the reciprocal productivity index method has been applied on the early time flowback data to evaluate the stimulated vertical gas wells (Crafton, 1996, Crafton, 1997 and Crafton, 1998). However, application of this approach for analyzing the flowback data of fractured horizontal wells needs further modifications. Ilk et al., 2010a and Ilk et al., 2010b introduced a workflow for a qualitative interpretation of early time flowback data by developing various diagnostic plots to observe wellbore unloading and fracture clean-up/depletion trends. Clarkson (2012) presented a quantitative analysis of two-phase flowback data using a two-phase tank model simulator to estimate fracture permeability and total fracture half length. Later, Clarkson's model was improved by applying Monte Carlo simulation for stochastic history matching of two-phase flowback data measured after multistage hydraulic fracturing (Williams-Kovacs and Clarkson, 2013). In addition to rate transient models, compositional simulators have been developed to history-matching flowback salt concentration change (Gdanski et al., 2007).

This paper aims at 1) Qualitative and careful analysis of multi-phase flowback data for understanding water displacement patterns, and 2) development of a simple analytical tool for analyzing early-time rate and pressure data. The second objective is achieved by extending the existing models of fracture testing. Various flow and shut-in tests have been proposed for recording the fracture response transferred by the fracturing fluid. Examples include the injection/fall off test (Craig, 2006), the fracture-calibration test (Mayerhofer et al., 1995) and the slug test (Peres et al., 1993). The mathematical models for such tests are developed by solving the material balance equation for fluid transport in the reservoir, fracture, and wellbore. The solutions have been reported in the form of type curves (Craig, 2006). The main out puts of the fracture tests are fracture conductivity and storativity.

The remaining of this paper is organized as follows: Section II qualitatively interprets the rate, pressure, and cumulative production of water and oil/gas recorded during three different flowback operations. Section III develops a simple analytical model to compare the pressure/rate transient behavior of the three flowback cases. Section IV applies the proposed model to the field data and discusses the results. SectionV discusses the overall results and summaries the paper.

FLOWBACK RATE AND PRESSURE HISTORY

In this section, we interpret flowback rate and pressure history of three multifractured horizontal wells completed in one tight oil and two tight gas reservoirs. Table 1 shows the completion data and fluid properties of the three wells.

Table 1: Completion data and fluid properties of three wells

Given parameters	Well 1	Well 2	Well 3
Hydrocarbon type	Gas	Gas	Oil
Fracturing fluid	Water	Water	Water
Distance between fracture stages (L_f), ft	242.78	91.86	236.22
Horizontal well length (X_e), ft	4593.17	1312.33	4265.09
Number of fracture stages (N_f)	20	15	20
Total compressibility (c_t), psi^{-1}	2.850 e^{-4}	2.871e^{-4}	2.901e^{-4}
Water compressibility (c_w), psi^{-1}	3.333e^{-6}	3.333e^{-6}	3.333e^{-6}
Viscosity of fracturing fluid (μ), cp	0.331	0.331	0.331
Water formation volume factor (B_w),	1.0311	1.0290	1.0003
Wellbore radius (r_w), ft	0.2916	0.2998	0.2874

Well 1

This well is completed in a tight gas reservoir. Initially, the well was flowed back with variable choke sizes for couple of hours. Then, two different choke sizes of 19.05 mm and 38.10 mm were used for almost 24 h and 48 h, respectively.

Flowback History

Fig. 1(a) shows the flow rate and pressure measured at the surface during the flowback of well 1. Casing pressure is initially high and quickly drops with time. The rate plot is divided into three regions.

In the first region, $q_g = 0$ and only water flows with a rate specified by the choke size. In the second region, gas production starts and q_w gradually decreases. In the third region, $q_w \approx 0$ and mainly gas is produced.

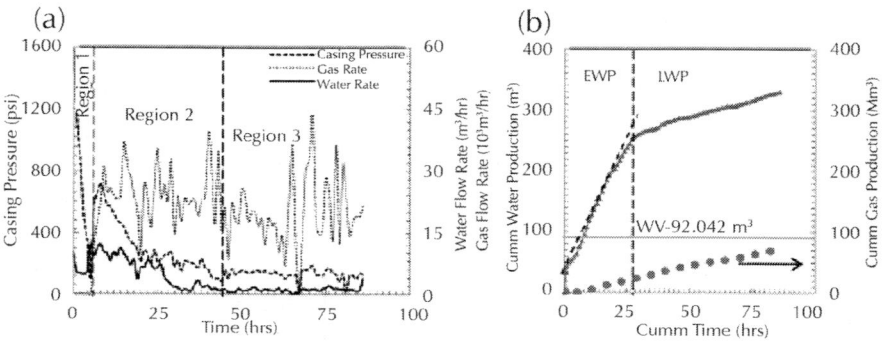

Figure 1: (a) Pressure and flow rate history measured every one hour during the flowback operation of well 1. Three different regions are identified. In region 1 water production dominates. In region 2 water production decreases and gas production increases. In region 3 gas production dominates. (b) Comparison between cumulative water and gas production curves and wellbore volume.

Fig. 1(b) compares the cumulative water production and gas production versus cumulative time. Cumulative water production curve shows two distinct regions. The first region is denoted by a black dashed line which shows a steep increase in water production for about 25 h and is named Early Water Production (EWP) region. The second region shows the gradual increase in water production until the end of flowback operation and is named Late Water Production (LWP) region. During EWP, water flow rate (determined by the curve slope) remains relatively high. During LWP, water flow rate decreases gradually. Faster initial water production rate can be explained by two reasons: 1) Water saturation and in turn, water relative permeability in fractures is initially high and drops with time as gas is introduced from the matrix into the fractures. 2) Initially conductive primary fractures contribute to water production, followed by secondary fractures with a relatively less conductivity. The gas production curve, in Fig. 1 (b), shows that gas breaks through almost 5 h after opening the well, and cumulative production gradually increases. This indicates gradual gas saturation increase or water saturation drop that was discussed above. Table 2 lists the relative volumes of water recovered during the flowback of this well. The total injected volume (TIV) is 1501 m³. After 86 h of flowback, the total load recovery (TLR) is 329.64 m³, which is only 21.96% of TIV. During EWP, 261 m³ of water is produced which is about 79.17% of TLR and the remaining 20.83% is recovered during LWP. The wellbore volume (WV) is 92.042 m³, which is initially filled with water and contributes to 27.92% of TLR.

Table 2: Comparison of relative volumes of water recovered during flow-back of the well 1, well 2 and well 3

Relative water volume	Well 1	Well 2	Well 3
Total injected volume (TIV), m³	1501	6443.38	2783.2
Breakthrough time, h	5	6	38
Total flowback time, h	86	116	75
Total load recovery (TLR), m³	329.64	1521.16	1346.51
Wellbore volume (WV), m³	92.042	74.089	103.922
Ratio of TLR to TIV, %	21.96	23.60	48.37
Ratio of WV to TLR, %	27.92	4.87	7.71
Ratio of LR @ EWP to TLR, %	79.17	49.96	89.11

Well 2

This well is completed in a tight gas reservoir. Initially, the well was flowed back with five choke sizes for 16 h. Then a choke size of 19.05 mm was used for 100 h.

Flowback History

Fig. 2(a) shows the flow rate and pressure measured at the surface during the flowback of Well 2. Tubing pressure is initially high and quickly drops with time. Several peaks followed by decline behaviors are observed in the pressure plot, which indicate that this well has been shut-in several times after starting the flowback operation. The rate and pressure plot is divided into three regions. In the first region, q_g is relatively low and water production dominates with a rate specified by the choke size. In the second region, gas flow rate ramps up and q_w gradually decreases in different steps, which are specified by the choke size. In the third region, q_w is relatively low and gas production dominates.

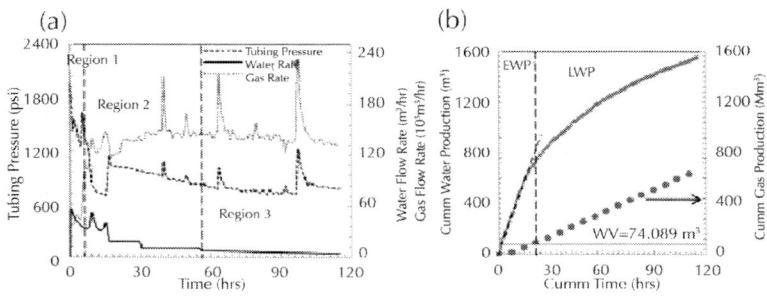

Figure 2: (a) Pressure and flow rate history measured every one hour during the flowback operation of well 2. Three different regions are identified. In region 1 water production dominates. In region 2 water production decreases and gas production increases. In region 3 gas production dominates. (b) Comparison between cumulative water and gas production curves and wellbore volume.

Fig. 2(b) compares the cumulative water and gas production versus time. Similar to well 1, the cumulative water production curve here shows two distinct regions. The first region (EWP) is denoted by a black dashed line which shows a steep increase in water production for about 24 h. The second region (LWP) shows a gradual increase in water production until the end of flowback operation. The relatively high water flow rate during EWP, and its gradual decrease during LWP can be explained by relative permeability effect as was done for well 1. The other curve in Fig. 2 (b) shows the gradual increase in gas production after the breakthrough that occurs almost 6 h after opening the well.

Table 2 shows that TIV for well 2 is 6443.38 m³. After 116 h of flowback 1521.16 m³ of water (TLR) is recovered that is only 23.60% of TIV. During EWP, 760 m³ of water is produced which is about 49.96% of TLR, and the remaining is recovered during LWP. The wellbore volume is 74.089 m³, which is initially filled with water and contributes to 4.87% of TLR.

Well 3

This well is completed in a tight oil reservoir. Initially, the well was flowed back with seven choke sizes for about 38 h. Then a choke size of 38.10 mm was used for about 37 h.

Flowback History

Fig. 3(a) shows the flow rate and pressure measured at the surface during the flowback of Well 3. Casing pressure is initially high and quickly drops with time. Similarly, the rate and pressure plot is divided into three regions. In the first region, q_o is zero and water production dominates with a rate specified by the choke size. In the second region, oil flow rate ramps up and q_w gradually decreases. In the third region, q_w is relatively low and oil production dominates.

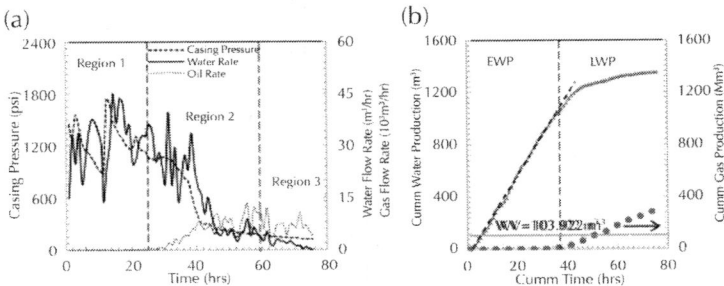

Figure 3: (a) Pressure and flow rate history measured every one hour during the flowback operation of well 3. Three different regions are identified. In region 1 water production dominates. In region 2 water production decreases and oil production increases. In region 3 oil production

dominates. (b) Comparison between cumulative water and oil production curves and wellbore volume.

Fig. 3(b) compares the cumulative water and oil production versus time. Again cumulative water curve shows two distinct regions. The first region (EWP) is denoted by a black dashed line which shows a steep increase in water production for about 38 h. During the second region (LWP) water production slowly increases and reaches to a constant value at the end of flowback operation. Interestingly, this plateau observed here was not observed in the previous two gas cases. Furthermore, the oil breakthrough occurs at a much later time compared with gas breakthrough in previous cases. Similarly, the fast water production during EWP can be explained by relative permeability effect. The other curve shows that 38 h after opening the well, oil breaks through and its production gradually increases. One should note that water recovery curve shown in Fig. 3(b) is analogous to oil recovery curve in water flood projects. After oil breakthrough, water cumulative curve deviates from the linear behavior, and water rate gradually decreases to very low values.

Table 2 shows that TIV in well 3 is 2783.2 m^3, and 1346.51 m^3 of that is recovered after 75 h of flowback operation. This means that flowback efficiency (TLR/TIV) is 48.37% that is more than two times of flowback efficiency for the previous two gas cases. This can be partly explained by the lower mobility of oil compared with gas that leads to a more efficient water displacement in fractures. This argument is backed with later breakthrough of oil compared with that of gas observed in the first two field cases. The wellbore volume (WV) here is 103.922 m^3, which is initially filled with water and contributes to 7.71% of the TLR. During EWP, 1200 m^3 of water is produced which is about 89.11% of TLR and the remaining recovered water is produced during LWP.

Comparative Interpretation

The rate plots of the three field cases consistently show three regions:

- Region 1, where water production dominates.
- Region 2, where water production drops and hydrocarbon production increases.
- Region 3, where hydrocarbon production dominates.

Region 1 is very short for the gas wells while it lasts much longer for the oil well. Furthermore, this region is influenced by wellbore storage. The data presented in Table 2, and Fig. 1, Fig. 2 and Fig. 3 indicate that the volume of water recovered during region 1 is comparable to the volume of wellbore. This is more pronounced for well 1 as is indicated in Fig. 1(b). After oil or gas breakthrough (region 2 and region 3) the phase saturation (S_w, S_o or S_g) in fractures change with time, and the system variables include

- Phase saturation (S_w, S_o or S_g)
- Phase mobility (k_w/μ_w, k_o/μ_o or k_g/μ_g)
- Total compressibility ($C_t = C_g S_g + C_o S_o + C_w S_w + C_m$)

We further classify the flowback history based on the water and gas/oil production curves into two major periods of EWP and LWP. Table 2 compares the relative volumes of water recovered during the flowback of the three wells. The low flowback efficiency of wells 1 and 2 compared with well 3 is consistent with early gas breakthrough compared with relatively late oil breakthrough due to its lower mobility. Gas can easily channel through water especially in vertical fractures below the horizontal well, that leads to a poor sweep efficiency (Parmar et al., 2012 and Parmar et al., 2013). This partly explains why the ratio of TLR to TIV is only 21.96% and 23.60% for wells 1 and 2 respectively, while it is 48.37% for well 3. Furthermore, in contrast to wells 1 and 2, the water recovery curve of well 3 reaches to a plateau that can be explained by a similar argument. We also observe that the fraction of TLR produced during EWP (early linear part of water production curve) for well 2 (49.96%) is much lower than that for wells 1 and 3 (79.17% and 89.11%, respectively). This indicates inefficient flowback of water for well 2 at early time scales possibly due to extensive water leak-off and/or two-phase (relative permeability) effects.

Approximate Volume and Interface of Fractures Depleted After Flowback

We can also estimate the depleted fracture volume by using the cumulative water recovery and a simple material balance. Assuming negligible water influx from the matrix, the recovered water volume is given by

$$V_{Rec} = Vf \phi f(1-Sw)$$

Here, V_{Rec}, ϕ_f and S_w, represent total water volume recovered, fracture porosity and water saturation left in the hydraulic fractures at that point in time, respectively. Therefore, depleted fracture volume V_f is given by

$$V_f = \frac{V_{Rec}}{\phi_f(1 - S_w)}$$

This equation is only valid if we assume that the water produced during the flowback comes from the induced fractures, and matrix water influx is negligible. One should note that this assumption does not mean that there is no water imbibition or leak off. Instead, it means that the imbibed or leaked-off water can hardly be produced back due to the capillarity and relative permeability effects. Dilution of fracture water with formation water or leak-off water can be investigated by flowback chemical analysis (Asadi et al., 2008), that is beyond the scope of this work. Table 3 shows the approximate fracture volume for the three flowback cases. ϕ_f and S_w are uncertain parameters, and are assumed to be 48% and 30%, respectively.

Table 3: Approximate depleted fracture volume $V_{Rec}/\phi_f(1-S_w)$ and matrix-fracture cross-sectional area created per stage $V_f/(W_f.N_f)$ at the end of the flowback operation of the three wells

		Well 1	Well 2	Well 3
V_f (ft³)		34646	159930	141520
w_f (m)	w_f (ft)	A_{cm} (ft²)	A_{cm} (ft²)	A_{cm} (ft²)

0.005	0.0164	105,601	650,000	431,000
0.003	0.0098	176,001.7	1,080,000	719,000
0.001	0.0033	528,005	3,250,000	2,160,000
0.0005	0.0016	1,056,010	6,500,000	4,310,000

We can also estimate the fracture interface created per stage by using the depleted fracture volume and assuming an average fracture aperture

$$A_{cm} = \frac{V_f}{\left(W_f \cdot N_f\right)}$$

Here, A_{cm}, W_f and N_f represent the matrix-fracture cross-sectional area, fracture aperture and number of fracture stages, respectively. Table 3 and Fig. 4 present the matrix-fracture cross-sectional area created per stage for different values of fracture aperture. We assume four sets of fracture aperture varying from 0.5 mm to 0.5 cm to estimate the matrix-fracture cross-sectional area created per stage.

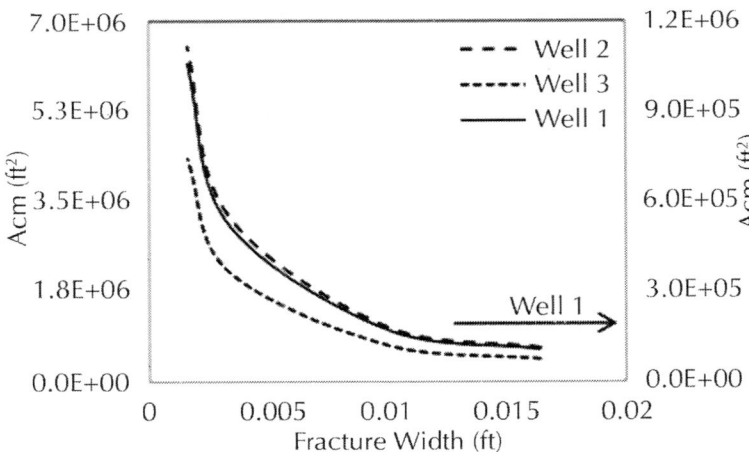

Figure 4: Average matrix-fracture interface created per stage versus fracture aperture at the end of the flowback operation for the three wells.

COMPARING THE PRESSURE TRANSIENT BEHAVIOR

This section develops an analytical model to compare the pressure transient behavior of the three wells. First, we describe the conceptual model assumed for developing the governing equations. Then we combine the solutions of continuity and diffusivity equations to present an analytical solution for the average fracture pressure. Finally, we develop a linear relationship between rate normalized pressure (RNP) and material balance time (MBT) by assuming negligible matrix influx at early time scales.

Conceptual Model

The conceptual model used in this study is similar to that proposed by Fan et al. (2010) and Clarkson (2012). Fig. 5 shows a multifractured horizontal well with multiple perforation clusters per stage.

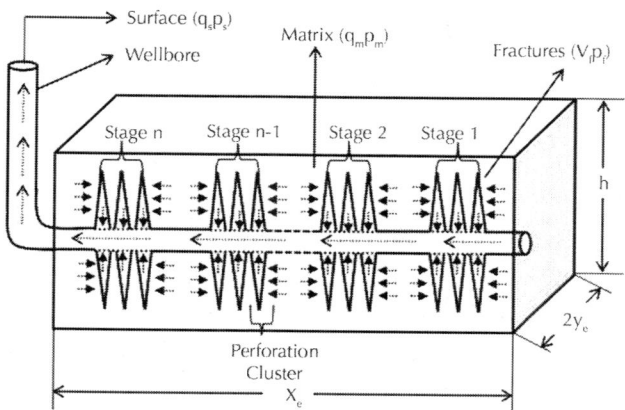

Figure 5: 3D view of a fractured horizontal well considered for developing the material balance equation. Dashed arrows show fluid flow direction, which is sequentially from matrix to fractures, fractures to wellbore, and finally from wellbore to surface.

A stimulated tight reservoir consists of 1) the wellbore (wb) consisting of the horizontal and vertical sections; 2) the fractures (f), which are mainly vertical if the minimum stress is in the horizontal direction; and 3) the rock matrix (m). The vertical hydraulic fractures are connected to the horizontal well with the length of X_e. The formation thickness and fractures half-length are denoted by h and y_e, respectively.

Material Balance Equation for Early Time Flowback

We use the mass conservation law to develop a relationship between the water production rate and average system pressure drop with respect to time. Fig. 5 shows the control volume, which includes the hydraulic fractures and wellbore including both horizontal and vertical sections. The material balance equation is given by

Mass in-Mass out=Accumulation   (1)

$$q_m \rho_m B_m - q_s \rho_s B_s = \frac{d}{dt}\left(\rho_f V_f + \rho_{wb} V_{wb}\right)$$

(2)

Here, subscripts f, s, wb and m denote fracture, surface, wellbore and matrix, respectively. B represents formation volume factor, which is the ratio between the fluid volume in the reservoir to that on the surface conditions.

We assume single phase flow at very early times. Therefore, we assume no matrix influx ($q_m = 0$) for that short time period. Expanding the derivative term on the right-hand side of Eq. (2) gives

$$-q_s \rho_s B_s = \rho_f \frac{d}{dt} V_f + V_f \frac{d}{dt} \rho_f + V_{wb} \frac{d}{dt} \rho_{wb}.$$

(3)

The first term on the right-hand side describes the change in fracture volume with time. This term is negative during the fracture closure, and it is zero after fracture closure. The second and third terms represent the change in the density of the fluid in fractures and wellbore, respectively. We further simplify the above equation by using the chain rule:

$$-q_s \rho_s B_s = \rho_f \frac{dV_f}{dP_f}\frac{dP_f}{dt} + V_f \rho_f \frac{1}{\rho_f}\frac{d\rho_f}{dP_f}\frac{dP_f}{dt} + V_{wb}\rho_{wb}\frac{1}{\rho_{wb}}\frac{d\rho_{wb}}{dP_{wb}}\frac{dP_{wb}}{dt}.$$
(4)

By considering the definition of isothermal compressibility for wellbore fluid (C_{wb}) and fracture fluid (C_f), the above equation can be written as

$$-q_s \rho_s B_s = \rho_f \frac{dV_f}{dP_f}\frac{dP_f}{dt} + V_f \rho_f C_f \frac{d\rho_f}{dt} + V_{wb}\rho_{wb}C_{wb}\frac{dP_{wb}}{dt}.$$
(5)

We assume that $\rho_s \approx \rho_{wb} \approx \rho_f$. This assumption means that the average density of fluid recovered at the surface, that of fluid in the wellbore, and in the fractures are almost equal.

We also assume $dP_{wb}/dt \approx dPf/dt \approx d\ \bar{P}/dt$, that means the rate of change of pressure with respect to time in the wellbore is almost equal to that in the fracture, and is given by an average pressure drop with respect to time in the control volume $d\ \bar{P}/dt$. Now Eq. (5) becomes

$$-q_s B_s = \left(\frac{dV_f}{dP_f} + V_f C_f + V_{wb}C_{wb}\right)\frac{d\bar{P}}{dt}.$$
(6)

The   total   storage   coefficient   is   defined

$$C_{st} = \frac{dV_f}{dP_f} + V_f C_f + V_{wb}C_{wb}.$$
(7)

The first term on the right-hand side accounts for fracture closure. The second and third terms represent the fracture and wellbore storage, respectively. One should note that the cumulative water production plots in Figs. 1(b), 2(b) and 3(b) and the estimations given in Table 3 indicate that $V_f \gg V_{wb}$. Finally, the material balance equation is given by

$$\frac{d\bar{P}}{dt} = \frac{-q_s B_s}{C_{st}}.$$

(8)

Combining Material Balance and Diffusivity Equations

We consider radial and linear flow of fracture water toward the horizontal well, as shown in Fig. 6. The fracture height and fracture half-length are denoted by h_f and y_e, respectively.

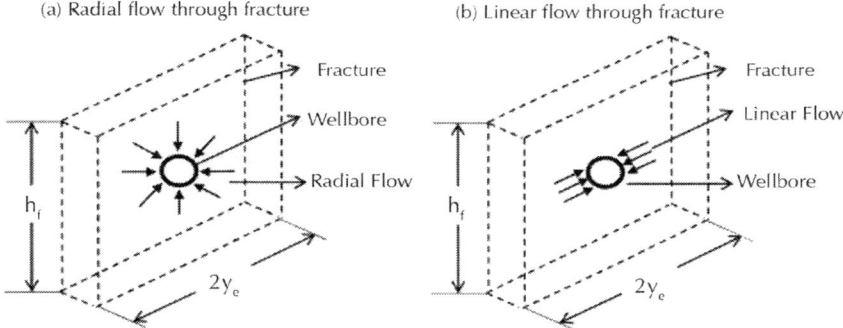

Figure 6: 3D view of a fracture with a horizontal well in the center considered for solving the diffusivity equation. Bold arrows show (a) radial and (b) linear water flow towards the horizontal well.

Radial Transient Model

The radial diffusivity equation for single-phase water flow through the hydraulic fractures towards the horizontal well is given by

$$\frac{1}{r}\frac{\partial}{\partial r}\left(r\frac{\partial P_f}{\partial r}\right) = \frac{\phi_f C_t \mu}{K_f}\frac{\partial P_f}{\partial t}.$$

(9)

The application of Eq. (9) also involves the following assumptions:

• Negligible gravity effect

- Constant temperature and viscosity
- Constant porosity, permeability, and total compressibility
- Negligible fluid influx from matrix into the fractures

Equation (9) can be solved under the following boundary conditions:

- $\partial P/\partial r = 0$ at $r = r_e$
- $P = P_{wf}$ at $r = r_w$
- $r_w^2/r_e^2 \approx 0$

Therefore, the fracture pressure in time and space is given by (see Appendix A).

$$P_f(r,t) = P_{wf} - \frac{\phi_f C_t \mu}{K_f} \frac{q_s B_s}{2C_{st}} r_e^2 \left[\ln\left(\frac{r_w}{r}\right) + \frac{r^2}{2r_e^2} \right]. \tag{10}$$

Where r_e is an equivalent fracture drainage radius. The average fracture pressure as a function of time is

$$\overline{P}(t) = P_{wf} - \frac{\phi_f C_t \mu}{K_f} \frac{q_s B_s}{2C_{st}} r_e^2 \left[\ln\left(\frac{r_w}{r_e}\right) + \frac{3}{4} \right]. \tag{11}$$

In reality, the fracture geometry is not circular. The following generalized solution applies for different fracture geometries (see Appendix A):

$$\overline{P}(t) = P_{wf} + \frac{\phi_f C_t \mu}{K_f} \frac{q_s B_s}{2C_{st}} r_e^2 \left[\frac{1}{2} \ln\left(\frac{4A}{C_A \gamma r_w^2}\right) \right]. \tag{12}$$

Where, A is the area of a single vertical fracture and C_A is the shape factor, which specifies the fracture geometry.

Linear Transient Model

A similar approach can be followed to solve the system pressure assuming linear flow of fracturing fluid towards the horizontal well (see Fig. 6(b)). The derivation details are given in Appendix B, and the final solution is given below that is analogous to Eq. (12).

$$\bar{P}(t) = P_{wf} + \frac{\phi_f C_t \mu}{K_f} \frac{q_s B_s}{3C_{st}} y_e^2.$$

$$(13)$$

Relationship between RNP and MBT

Fluid expansion and fracture closure are the dominant mechanisms at early time scales in the absence of matrix influx (qm=0):

$$N_P B_s = -C_{st}(\bar{P} - P_i).$$

$$(14)$$

Where, NP is the cumulative fracturing fluid production. Rearranging Eq. (14) gives the average fracture pressure:

$$\bar{P}(t) = P_i - \frac{N_P B_s}{C_{st}}.$$

$$(15)$$

Substituting Eq. (15) into Eq. (12), and dividing both sides by $q_s B_s$ gives

$$\frac{P_i - P_{wf}}{q_s} = \frac{N_P B_s}{q_s C_{st}} + \frac{\phi_f C_t \mu B_s}{2C_{st}K_f} r_e^2 \left[\frac{1}{2} \ln\left(\frac{4A}{C_A \gamma r_w^2} \right) \right].$$

$$(16)$$

The terms $(P_i - P_{wf})/q_s$ and N_p/q_s are referred to as rate normalized pressure (RNP) and material balance time (MBT):

$$\text{RNP} = \frac{B_s}{C_{st}} \text{MBT} + \frac{\phi_f C_t \mu B_s}{2C_{st}K_f} r_e^2 \left[\frac{1}{2} \ln\left(\frac{4A}{C_A \gamma r_w^2} \right) \right].$$

$$(17)$$

An analogous expression can be derived for linear flow starting from Eq. (13):

$$\text{RNP} = \frac{B_s}{C_{st}} \text{MBT} + \frac{\phi_f C_t \mu B_s}{3C_{st}K_f} y_e^2.$$

$$(18)$$

Equations (17) and (18) describe a linear relationship between RNP and MBT. These equations are analogous to the equations proposed by Palacio and Blasingame (1993) for application of material balance time for boundary dominated liquid and gas flow in vertical wells. The proposed equations are also analogous to the flowing material balance equation (FMB) (Agarwal et al., 1999, Mattar and Anderson, 2003 and Mattar and Anderson, 2005). Clarkson et al. (2008) demonstrated that FMB could be applied to single phase coal bed methane (CBM) reservoirs. Furthermore, application of MBT and RNP has been recently discussed and applied for production data analysis of shale gas reservoirs (Song et al., 2011; andSong and Ehlig-Economides, 2011). Equations (17) and (18) can be used for history matching the production data measured during early-time flowback with a relatively high frequency and accuracy. The line slope can be interpreted to estimate the total storage coefficient defined by Eq. (7). If other parameters are known, the intercept can be used to characterize the fracture geometry by calculating (A/C_A) for the radial case, and fracture half length (ye) for the linear case. However, successful application of this model requires high frequency and accurate rate and pressure data.

MODEL APPLICATION

The proposed model can be used to history match single phase water rate and pressure measured at the beginning of flowback operation. Therefore, region 1 provides the most representative data set for history matching using Eqs. (17) and (18).

Analysis Procedure

We propose the following analysis procedure:
- Obtain early-time flowback pressure and rate data.
- Plot rate normalized pressure (RNP) versus material balance time (MBT).
- Determine the slope and intercept of the best linear match.
- Calculate total storage coefficient (C_{st}) by using the line slope.
- Obtain a relationship for dimensionless radial fracture parameter $(\phi_f K_f) r_e^2 \left[1 / 2\ln\left(4A / C_A \gamma r_w^2 \right) \right]$ and dimensionless

linear fracture parameter $\left(\phi_f / K_f\right) y_e^2$ by using line intercept and C_{st} determined in step 4.

Example Applications

Unfortunately, the pressure and rate data are not measured with sufficient frequency required for an accurate analysis. Furthermore, we observe an early gas breakthrough for the first two cases (wells 1 and 2) that shortens the duration of region 1 described by the proposed model. However, we find it useful to demonstrate the application of the proposed model by using region 1 of the three field data sets. Table 1shows the completion data and fluid properties of the three wells. We first plot rate normalized pressure (RNP) versus material balance time (MBT) for region 1 of the three wells as shown in Fig. 7. A linear relationship in the form of *RNP = m MBT + b* is obtained in each case, where *m* and *b* can be interpreted as

$$m = \frac{B_S}{C_{st}}$$

$$b = \frac{\phi_f C_t \mu B_S}{2C_{st}K_f}r_e^2\left[\frac{1}{2}\ln\left(\frac{4A}{C_A\gamma r_w^2}\right)\right] \text{(for radial fracture depletion)}$$

$$b = \frac{\phi_f C_t \mu B_S}{3C_{st}K_f}y_e^2 \quad \text{(for linear fracture depletion)}$$

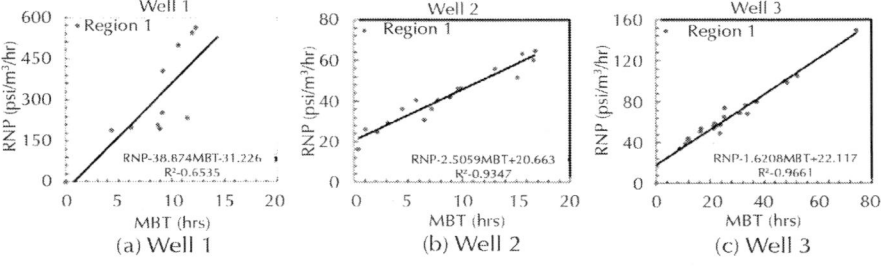

Figure 7: RNP versus MBT of Region 1 and the best linear fit for the three wells.

The line intercept for well 1 is negative that can't be described by the proposed model. The data points in this case are scattered. Dominance of wellbore volume and early gas breakthrough are among several reasons responsible for the lack of match. A better linear match is observed for the other two wells. We first use the line slope to calculate the total storage coefficient (C_{st}) for the three multifractured horizontal wells. Next, we use the line intercept and other known parameters to obtain a relationship for the dimensionless fracture parameters; $(\phi_f K_f) r_e^2 \left[1 / 2\ln\left(4A / C_A \gamma r_w^2\right)\right]$ for the radial case, and $(\phi_f / K_f) y_e^2$ for the linear case. Table 4 lists the calculated values of total storage coefficient and the dimensionless fracture parameters for the three cases.

Table 4: Calculated values for the total storage coefficient (C_{st}), the dimensionless radial fracture parameter $(\phi_f K_f) r_e^2 \left[1 / 2\ln\left(4A / C_A \gamma r_w^2\right)\right]$ and the dimensionless linear fracture parameter $(\phi_f / K_f) y_e^2$

Calculated parameters	Well 1	Well 2	Well 3
Slope (*m*)	38.874	2.5059	1.6208
Intercept (*b*)	−31.226	20.663	22.117
Total storage coefficient (*Cst*), ft³/psi	0.936	14.504	21.800
$\frac{\phi_f}{K_f} r_e^2 \left[\frac{1}{2}\ln\left(\frac{4A}{C_A \gamma r_w^2}\right)\right]$, ft²/md	N/A	45.759	74.942
$\frac{\phi_f}{K_f} y_e^2$, ft²/md	N/A	68.638	112.414

Discussion of Results

The transient analysis leads to the following key observations:

- The negative line intercept for well 1 can not be described by the proposed model.

- Total storage coefficient of well 3 is almost 30% higher than that of well 2.
- The dimensionless fracture parameter of well 3 is higher than that of well 2 for both radial and linear cases.

Result 1 indicates that the proposed model requires high frequency pressure and rate measurement before gas breakthrough. Furthermore, the wellbore volume should be relatively low enough compared with water volume produced before gas breakthrough.

Result 2 can be explained by comparing the completion and stimulation of wells 2 and 3: (i) The wellbore volume of well 3 is 30% higher than that of well 2. (ii) The number of fracture stages in well 3 is 25% higher than that in well 2. (iii) The water volume injected per stage in well 3 is almost 30% of that in well 2. In contrast to the first two items, item (iii) is not in agreement with the observed trend. Although a lot more water is injected for treatment of well 2, it does not lead to a higher storage coefficient based on the proposed analysis. This is backed with the observation that flowback efficiency of well 3 is more than two times of that of well 2. Furthermore, a large volume of water injected in well 2 can leak off into the gas saturated matrix, and does not contribute to fracture storage. In addition, some of the induced hydraulic fractures may be cut-off from the effective hydraulic fracture network and in turn can not contribute to the flow. Hence some of the water becomes trapped in the ineffective hydraulic fracture clusters.

Result 3 indicates that the fracture length scale for well 3 is higher than that for well 2. Assuming equal fracture porosity and permeability, the fracture half length of well 3 is estimated to be 20% higher than that of well 2. Therefore, it qualitatively complements result 2 that indicates higher induced fracture volume of well 3 compared with that of well 2. Results 2 and 3 indicate that fracturing operation of well 3 is more successful than that of well 2. However, the amount of water used for treatment of each stage in well 2 is almost three times higher than that in well 3. This can be explained by a stronger water leak-off in well 2 that is a gas well compared with well 3 that is an oil well.

DISCUSSION AND SUMMARY

We compared the flowback behavior of three multifractured horizontal wells by conducting volumetric and pressure/rate transient analysis.

Volumetric Analysis

In general a consistent behavior is observed when plotting the cumulative production of water and gas/oil measured during the three flowback operations. The cumulative production plots of the three cases demonstrate two dominant regions. Initially water production linearly increases with time, and the main part of recovered water is produced during this period. After gas/oil breakthrough the cumulative curve deviates from the linear behavior, and for the oil case it reaches to a plateau. The flowback efficiency (total load recovery divided by total injected volume) of the oil case is more than two times of the gas cases. This can be explained by 1) unfavorable mobility ratio and gravity effects for displacement of water by gas (Parmar et al., 2012 and Parmar et al., 2013), and 2) forced and spontaneous imbibition of water into the gas saturated rock matrix (Dehghanpour et al., 2012 and Dehghanpour et al., 2013). The first explanation is backed by much earlier breakthrough of gas compared with that of oil observed here.

Transient Analysis

The rate plots of the three wells consistently show three regions. In region 1, water phase production dominates, while in region 3, gas/oil production dominates. Region 2 is the transition period where water rate decreases quickly and gas/oil rate increases. The length of region 1 for the oil well is significantly longer than that for the gas wells due to the quick gas breakthrough as discussed above. The simple analytical model presented in this paper that can be considered as an extension of previous models (Crafton, 1996,

Crafton, 1997 and Crafton, 1998) can be used for the analysis of region 1 data. The main out puts of this model are total storage coefficient and dimensionless fracture parameter that can be interpreted to obtain the fracture half-length. However, it should be noted that application of the proposed model, similar to that of previous single-phase models, requires high frequency pressure and rate data measured at the beginning of flowback operation. We used the proposed model to describe the flowback transient behavior of the three wells studied here. Although, the measured data do not have the sufficient frequency, the estimated values of total storage coefficient and dimensionless fracture parameter are in agreement with the results of volumetric analysis.

Analysis of regions 2 and 3 by analytical or semi-analytical methods is challenging since it requires simultaneous solution of 1) transient multi-phase flow in the wellbore, 2) Fracture/reservoir water saturation change in time and space, and 3) bottomhole multi-phase pressure and rate. However, there have been recent attempts to solve this problem by semi-analytical, simulation and stochastic techniques (Clarkson, 2012 and Williams-Kovacs and Clarkson, 2013; Ezulike et al., 2013).

Flowback Analysis of Shale Gas Wells

In this paper we analyzed the flowback data of tight oil and tight gas wells. Application of the proposed model on multifractured horizontal wells, completed in shales, remains the subject of a future study. Recently, Abbasi (2013) and Ghanbari et al. (2013) studied the flowback data of several multifractured horizontal wells, completed in the Horn River shales, by constructing various diagnostic plots. These studies demonstrate two dominant differences between the flowback behavior of the shale gas and tight gas (oil) wells:

- An immediate gas breakthrough is observed in shale gas wells, while in tight oil and tight gas wells, there is an initial period of single phase water flow.
- The plot of gas-water ratio (GWR) versus cumulative gas

production (G_p), for shale gas wells, shows two distinct regions. In the first region, GWR decreases with G_p, and in the second region it increases with G_p. However, the GWR plot of tight gas wells or OWR (oil-water ratio) of tight oil wells does not show the first region.

These two observations suggest that the fracture network of shale gas wells, before the flowback operation, is partly saturated with some initial gas. This initial free gas is possibly released during the shut-in period due to counter-current imbibition phenomenon (Dehghanpour et al., 2012, Dehghanpour et al., 2013 and Makhanov, 2013) and/or redistribution of the gas initially existing in the natural fracture network (Ghanbari et al., 2013). To further justify the different behaviors of the tight oil/gas and shale gas wells, Ezulike et al. (2013) compared the relative permeability versus time and relative permeability versus cumulative gas/oil production plots of the similar well groups. Consistently, they observed different relative permeability profiles for tight oil/gas and shale gas wells.

Data Acquisition

Accurate and frequent measurement of flow rate and pressure during flowback operations is critical for history-matching using the proposed models. Therefore, data collection during flowback operations is the first and most important step for flowback analysis. However, due to the operational issues, the data become noisy that may lead to discrepancies in the final analysis. Generally, flowback rate and pressure data are measured on hourly basis which is not sufficient for studying the transport phenomena quickly occurring at the early-time scales. In general, the flowing pressure can be measured with a high frequency (points/minute or points/second) (Ilk et al., 2010a and Ilk et al., 2010b), while the flow rate can not be measured with this frequency due to the limitations of flow rate measurement devices. However, the frequency of flow rate data can be improved by using the pressure data, cumulative production data, and a technique called wavelet analysis which reduces the uncertainty and noise from the data. Athichanagorn et al. (2002), Kikani and He (1998) and Ouyang and Kikani (2002) introduced the wavelet analysis technique for analysis of the data measured by permanent downhole gauges. Furthermore, it is strongly recommended to conduct a careful flow-regime analysis by constructing various diagnostic plots (Ilk et al., 2010a, Ilk et al., 2010b and Abbasi, 2013) before history-matching using the mathematical models.

Appendix A

In this section, we develop a relationship for fracture pressure in time and space. We use the well-known diffusivity equation which describe the radial flow of fluid in porous media (Lee et al., 2003). The diffusivity equation for radial flow of fracturing fluid through the hydraulic fractures towards the horizontal well is given by

$$\frac{1}{r}\frac{\partial}{\partial r}\left(r\frac{\partial P_f}{\partial r}\right) = \frac{\phi_f C_t \mu}{K_f}\frac{\partial P_f}{\partial t}.$$

(A.1)

Substituting Eq. (8) into Eq. (A.1) gives

$$\frac{1}{r}\frac{\partial}{\partial r}\left(r\frac{\partial P_f}{\partial r}\right) = \frac{\phi_f C_t \mu}{K_f}\frac{(-q_s)B_s}{C_{st}}.$$

(A.2)

Integrating with respect to r gives

$$\left(r\frac{\partial P_f}{\partial r}\right) = -\frac{\phi_f C_t \mu}{K_f}\frac{q_s B_s}{C_{st}}\frac{r^2}{2} + C_1.$$

(A.3)

We assume $\partial P/\partial r = 0$ at $r = r_e$ to determine C_1:

$$C_1 = \left(\frac{\phi_f C_t \mu}{K_f}\frac{q_s B_s}{2C_{st}}r_e^2\right).$$

(A.4)

Substituting C_1 in Eq. (A.3) gives

$$\frac{\partial P_f}{\partial r} = \frac{\phi_f C_t \mu}{K_f}\frac{q_s B_s}{2C_{st}}\left(-r + \frac{r_e^2}{r}\right).$$

(A.5)

Integrating with respect to r gives

$$P_f(r,t) = \frac{\phi_f C_t \mu}{K_f} \frac{q_s B_s}{2C_{st}} r_e^2 \left[-\frac{r^2}{2r_e^2} + \ln(r) \right] + C_2.$$
(A.6)

We consider the boundary condition $P = P_{wf}$ at $r = r_w$ and assume $r_w^2/r_e^2 \approx 0$ to calculate C_2:

$$C_2 = P_{wf} - \frac{\phi_f C_t \mu}{K_f} \frac{q_s B_s}{2C_{st}} r_e^2 \ln(r_w).$$
(A.7)

Substituting C_2 in Eq. (A.6) gives fracture pressure in time and space:

$$P_f(r,t) = P_{wf} - \frac{\phi_f C_t \mu}{K_f} \frac{q_s B_s}{2C_{st}} r_e^2 \left[\ln\left(\frac{r_w}{r}\right) + \frac{r^2}{2r_e^2} \right]$$
(A.8)

Now we develop a relationship between average reservoir pressure and bottomhole flowing pressure. Starting with the average reservoir pressure equation

$$\overline{P}(t) = \frac{\displaystyle\int_{r_w}^{r_e} P_f \cdot dV_f}{\displaystyle\int_{r_w}^{r_e} dV_f}.$$
(A.9)

We   assume   cylindrical   geometry

$$V_f = \pi r^2 h_f \phi_f.$$

$$dV_f = \pi h_f \phi_f 2r dr.$$

Substitute above equations in Eq. (A.9)

$$\overline{P} = \frac{\displaystyle\int_{r_w}^{r_e} \left(P_f \pi h_f \phi_f 2r\right)dr}{\displaystyle\int_{r_w}^{r_e} \left(\pi h_f \phi_f 2r\right)dr}.$$

$$\overline{P} = \frac{2\pi h_f \phi_f \displaystyle\int_{r_w}^{r_e} \left(P_f r\right)dr}{2\pi h_f \phi_f \displaystyle\int_{r_w}^{r_e} (r)dr}.$$

Here,

$$P_f(r,t) = P_{wf} - \frac{\phi_f C_t \mu}{K_f}\frac{q_s B_s}{2C_{st}}r_e^2\left[\ln\left(\frac{r_w}{r}\right) + \frac{r^2}{2r_e^2}\right]$$

Substituting Pf and solving

$$\overline{P} = \frac{1}{\left(\frac{r_e^2 - r_w^2}{2}\right)}\left\{\int_{r_w}^{r_e}\left[P_{wf} - \frac{\phi_f C_t \mu}{K_f}\frac{q_s B_s}{2C_{st}}r_e^2\left[\ln\left(\frac{r_w}{r}\right) - \frac{r^2}{2r_e^2}\right]\right]\right\}rdr.$$

(A.10)

Solving each part separately

$$P_{wf}\int_{r_w}^{r_e} rdr = P_{wf}\frac{r_e^2 - r_w^2}{2}.$$

Now solving second part

$$\int_{r_w}^{r_e} r\ln\left(\frac{r_w}{r}\right)dr.$$

Solving through Integration by Parts

$$\int u\cdot v' = u\cdot v - \int v\cdot u'.$$

Therefore the second part will become

$$-\frac{\phi_f C_t \mu (q_s) B_s}{K_f} \frac{}{2C_{st}} r_e^2 \int_{r_w}^{r_e} r \ln\left(\frac{r}{r_w}\right) dr = -\frac{\phi_f C_t \mu q_s B_s}{K_f} \frac{}{2C_{st}} r_e^2 \left[\frac{r_e^2}{2} \ln\left(\frac{r_w}{r_e}\right)\right.$$

$$\left. -\frac{1}{4}\left(r_e^2 - r_w^2\right)\right].$$

Solving the third part

$$\int_{r_w}^{r_e} \frac{r^3}{2r_e^2} dr = \frac{1}{2r_e^2}\left(\frac{r_e^4}{4} - \frac{r_w^4}{4}\right).$$

The final form of Eq. (A.9) becomes

$$\overline{P} = P_{wf} - \frac{\phi_f C_t \mu\, q_s B_s}{K_f} \frac{}{2C_{st}} r_e^2 \left[\ln\left(\frac{r_w}{r_e}\right)\left(\frac{r_e^2}{r_e^2 - r_w^2}\right) + \left(\frac{1}{2}\right)\right.$$

$$\left. +\frac{1}{4}\left(\frac{r_e^2 + r_w^2}{r_e^2}\right)\right].$$

Assuming that $\left(r_e^2/r_e^2 - r_w^2\right) \approx 1$ and $\left(r_e^2 + r_w^2/r_e^2\right) \approx 1$

$$\overline{P} = P_{wf} - \frac{\phi_f C_t \mu\, q_s B_s}{K_f} \frac{}{2C_{st}} r_e^2 \left[\ln\left(\frac{r_w}{r_e}\right) + \left(\frac{1}{2}\right) + \frac{1}{4}\right].$$

$$\overline{P} = P_{wf} - \frac{\phi_f C_t \mu\, q_s B_s}{K_f} \frac{}{2C_{st}} r_e^2 \left[\ln\left(\frac{r_w}{r_e}\right) + \frac{3}{4}\right].$$

$$(A.11)$$

Solving for various drainage shape

$$\ln\left(\frac{r_w}{r_e}\right) + \frac{3}{4}.$$

$$\ln\left(\frac{r_e}{r_w}\right) - \frac{3}{4} = \ln\left(\frac{r_e}{r_w}\right) + \ln e^{-3/4}.$$

$$\ln\left(\frac{r_e}{r_w}\right) - \frac{3}{4} = \ln\left(\frac{r_e}{r_w \cdot e^{3/4}}\right).$$

After solving Eq. (A.11) becomes

$$\bar{P}(t) = P_{wf} + \frac{\phi_f C_t \mu\, q_s B_s}{K_f}\frac{1}{2C_{st}}r_e^2\left[\frac{1}{2}\ln\left(\frac{4A}{C_A \gamma r_w^2}\right)\right]$$

(A.12)

Appendix B

In this section, we develop a relationship for fracture pressure in time and space. The diffusivity equation for linear flow of fracturing fluid through the hydraulic fractures towards the horizontal well is given by

$$\frac{\partial^2 P_f}{\partial x^2} = \frac{\phi_f C_t \mu}{K_f}\frac{\partial P_f}{\partial t}.$$

(B.1)

Substituting Eq. (8) into Eq. (B.1) gives

$$\frac{\partial^2 P_f}{\partial x^2} = \frac{\phi_f C_t \mu}{K_f}\frac{(-q_s)B_s}{C_{st}}.$$

(B.2)

Integrating with respect to x gives

$$\left(\frac{\partial P_f}{\partial x}\right) = -\frac{\phi_f C_t \mu}{K_f}\frac{q_s B_s}{C_{st}}x + C_1.$$

(B.3)

We assume $\partial P/\partial r = 0$ at $x = y_e$ to determine $C1$:

$$C_1 = \left(\frac{\phi_f C_t \mu}{K_f}\frac{q_s B_s}{C_{st}}y_e\right).$$

(B.4)

Substituting C_1 in Eq. (B.3) gives

$$\frac{\partial P_f}{\partial x} = \frac{\phi_f C_t \mu}{K_f}\frac{q_s B_s}{C_{st}}(-x + y_e).$$

(B.5)

Integrating with respect to x gives

$$P_f(x,t) = \frac{\phi_f C_t \mu}{K_f}\frac{q_s B_s}{C_{st}}\left[-\frac{x^2}{2} + y_e x\right] + C_2$$

(B.6)

We consider the boundary condition $P = P_{wf}$ at $x = 0$ to calculate C_2:

$$C_2 = P_{wf}$$

(B.7)

Substituting C_2 in Eq. (B.6) gives fracture pressure in time and space:

$$P_f(x,t) = P_{wf} - \frac{\phi_f C_t \mu}{K_f} \frac{q_s B_s}{C_{st}} r_e^2 \left[\frac{x^2}{2} - y_e x \right]$$

(B.8)

Now we develop the relationship between average reservoir pressure and bottomhole flowing pressure. Starting with the average reservoir pressure equation

$$\overline{P}(t) = \frac{\displaystyle\int_0^{y_e} P_f \cdot dV_f}{\displaystyle\int_0^{y_e} dV_f}.$$

(B.9)

We assume cuboid geometry

V=whx

$$V_f = w_f h_f \phi_f x.$$

$$dV_f = w_f h_f \phi_f dx.$$

Substitute above equations in Eq. (B.9)

$$\overline{P} = \frac{\displaystyle\int_0^{y_e} (P_f w_f h_f \phi_f) dx}{\displaystyle\int_0^{y_e} w_f h_f \phi_f dx}.$$

$$P_f(x, t) = P_{wf} - \frac{\phi_f C_t \mu}{K_f} \frac{q_s B_s}{C_{st}} \left[\frac{x^2}{2} - y_e x \right]$$

Here,

Substituting P_f and solving

$$\overline{P} = \frac{1}{y_e} \left\{ \int_0^{y_e} \left[P_{wf} - \frac{\phi_f C_t \mu}{K_f} \frac{q_s B_s}{C_{st}} \left[\frac{x^2}{2} - y_e x \right] \right] \right\} dx.$$

(B.10)

Solving for average pressure drop. The final form of Eq. (B.9) becomes

$$\overline{P} = P_{wf} + \frac{\phi_f C_t \mu}{K_f} \frac{q_s B_s}{3 C_{st}} y_e^2.$$

(B.11)

Relationship between RNP and MBT

As we assume that at very early time matrix influx is zero ($q_m = 0$), the production mechanism is fluid expansion and fracture closure:

$$N_P B_s = -C_{st} \left(\overline{P} - P_i \right).$$

(B.12)

Where, NP is the cumulative production of the fracturing fluid. Rearranging Eq. (B.12) gives the average fracture pressure as,

$$\overline{P}(t) = P_i - \frac{N_P B_s}{C_{st}}.$$

(B.13)

Substituting Eq. (B.13) into Eq. (B.11) gives

$$P_i - P_{wf} = \frac{N_P B_s}{C_{st}} + \frac{\phi_f C_t \mu}{K_f} \frac{q_s B_s}{3 C_{st}} y_e^2.$$

(B.14)

Dividing both sides of Eq. (B.14) by $q_s B_s$ gives

$$\frac{P_i - P_{wf}}{q_s} = \frac{N_P B_s}{q_s C_{st}} + \frac{\phi_f C_t \mu}{K_f} \frac{q_s B_s}{3 C_{st}} y_e^2.$$

(B.15)

The terms $(P_i - P_{wf})/q_s$ and N_P/q_s are referred to as rate normalized pressure (RNP) and material balance time (MBT).

$$\text{RNP} = \frac{B_s}{C_{st}} MBT + \frac{\phi_f C_t \mu B_s}{3 C_{st} K_f} y_e^2 .$$

(B.16)

ACKNOWLEDGMENTS

We thank FMC Technologies, Trican Well Service, and Natural Sciences and Engineering Research Council of Canada (NSERC) for financial support, and Lightsteam Resources for providing the flowback data.

REFERENCES

1. Abbasi, M.A., September 2013. A Comparative Study of Flowback Rate and Pressure Transient Behaviour in Multifractured Horizontal Wells (Master's thesis). University of Alberta, Edmonton, Alberta, Canada.

2. Abdelaziz, K., Hani, Q., Naiem, B., 2011. Tight gas sands development is critical to future world resources. In: Paper SPE 142049 Presented at the SPE Middle East Unconventional Gas Conference and Exhibition, Muscat, Oman, 31st January-2nd February.

3. Agarwal, R.G., Gardner, D.C., Kleinsteiber, S.W., Fussell, D.D., 1999. Analyzing well production data using combined-type-curve and decline-curve concepts. SPE Reserv. Eval. Eng. 2 (5), 478e486. http://dx.doi.org/10.2118/57916-PA. SPE- 57916-PA.

4. Asadi, M., Woodroof, R.A., Himes, R.E., May 2008. Comparative study of flowback analysis using polymer concentrations and fracturing-fluid tracer methods: a field study. SPE Prod. Oper. 23 (2), 147-157.

5. Athichanagorn, S., Horne, R.N., Kikani, J., 2002. Processing and interpretation of long-term data acquired from permanent

pressure gauges. SPE Reserv. Eval. Eng. 5 (5), 384e391. http:// dx.doi.org/10.2118/80287-PA. SPE-80287-PA.

6.   Bello, R.O., 2009. Rate Transient Analysis in Shale Gas Reservoirs with Transient Linear Behavior (Ph.D thesis). Texas A & M University, Texas.

7.   Clarkson, C.R., 2012. Modeling 2-phase flowback of multi-fractured horizontal wells completed in shale. In: Paper SPE 162593 Presented at the SPE Canadian Unconventional Resources Conference, Calgary, Alberta, Canada, 30th October-1st November.

8.   Clarkson, C.R., Jordan, C.L., Seidle, J.P., Gierhart, R.R., 2008. Production data analysis of coalbed-methane wells. SPE Reserv. Eval. Eng. 11 (2), 311e325. http:// dx.doi. org/10.2118/107705-PA. SPE-107705-PA.

9.   Crafton, J.W., 1996. The reciprocal productivity index method, a graphical well performance simulator. In: Paper Presented at the 43th Southwestern Petroleum Short Course, Lubbock, Texas, USA, 16 April.

10.   Crafton, J.W., 1997. Oil and gas well evaluation using the reciprocal productivity index method. In: Paper SPE 37409 Presented at the SPE Production Operations Symposium, Oklahoma City, Oklahoma, 9-11 March.

11.   Crafton, J.W., 1998.Well evaluation using early time post-stimulation flowback data. In: Paper SPE 49223 Presented at the SPE Annual Technical Conference and Exhibition, New Orleans, Louisiana, 27-30 September.

12.   Craig, D.P., 2006. Analytical Modeling of a Fracture-injection/ fall off Sequence and the Development of a Refracture Candidate Diagnostic Test (Ph.D thesis). Texas A & M University, Texas.

13.   Dehghanpour, H., Zubair, H.A., Chhabra, A., Ullah, A., 2012. Liquid intake of organic shales. Energy Fuels 2012 (26), 5750e5758. http://dx.doi.org/10.1021/ ef300979447.

14.   Dehghanpour, H., Lan, Q., Saeed, Y., Fei, H., Qi, Z., 2013. Spontaneous imbibition of brine and oil in gas shales: effect

of water adsorption and resulting microfractures. Energy Fuels 2013 (27), 3039e3049. http://dx.doi.org/10.1021/ef4002814.

15. El-Banbi, A.H., 1998. Analysis of Tight GasWell Performance (Ph.D thesis). Texas A & M University, Texas.

16. Ezulike, D.O., Dehghanpour, H., Hawkes, R.V., 2013. Understanding flowback as a transient 2-phase displacement process: an extension of the linear dual porosity model. In: Paper SPE 167164-MS Presented at the SPE Canadian Unconventional Resources Conference, Calgary, Alberta, Canada.

17. Fan, L., Thompson, J.W., Robinson, J.R., 2010. Understanding gas production mechanism and effectiveness of well stimulated in the Haynesville shale through reservoir simulation. In: Paper CSUG/SPE 136696 Presented at the SPE Unconventional Resources & International Petroleum Conference, Calgary, 19-21 October.

18. Gdanski, R., Weaver, J., Slabaugh, B., 29e31 January 2007. A new model for matching fracturing fluid flowback composition. In: Paper SPE 106040 Presented at SPE Hydraulic Fracturing Technology Conference, College Station, Texas, USA.

19. Ghanbari, E., Abbasi, M.A., Dehghanpour, H., Bearinger, D., 2013. Flowback volumetric and chemical analysis for evaluating load recovery and its impact on early-time production. In: Paper SPE-167165 presented at the SPE Canadian Unconventional Resources Conference, Calgary, Alberta, 5–7 November. http:// dx.doi.org/10.2118/167165-MS.

20. Ilk, D., Currie, S.M., Symmons, D., Rushing, J.A., Broussard, N.J., Blasingame, T.A., 2010a. A comprehensive workflow for early analysis and interpretation of flowback data from wells in tight gas/shale reservoir systems. In: Paper SPE 135607 Presented at the SPE Annual Technical Conference and Exhibition, Florence, Italy, 19e22 September.

21. Ilk, D., Anderson, D.M., Stotts, G.W.J., Mattar, L., Blasingame,

T.A., 2010b. Productiondata analysis, pitfalls, diagnostics. SPE Reserv. Eval. Eng. 13 (3), 538e552. http:// dx.doi. org/10.2118/102048-PA.

22. Kikani, J., He, M., 1998. Multiresolution analysis of long-term pressure transient data using wavelet methods. In: Paper SPE 48966 Presented at the SPE Annual Technical Conference and Exhibition, New Orleans, 27 -30 September. http:// dx.doi. org/10.2118/48966-MS.

23. Lee, J., Rollins, J.B., Spivey, J.P., 2003. Pressure Transient Testing. In: SPE Text Book Series, vol. 9.

24. Makhanov, K.K., 2013. An Experimental Study of Spontaneous Imbibition in Horn River Shales (Master's thesis). University of Alberta, Edmonton, Alberta, Canada.

25. Mattar, L., Anderson, D.M., 2003. A systematic and comprehensive methodology for advanced analysis of production data. In: Paper SPE 84472 Presented at the SPE Annual Technical Conference and Exhibition, Denver, 5e8 October.

26. Mattar, L., Anderson, D.M., 2005. Dynamic material balance (oil or gas-in-place without shut-ins). In: Paper 2005e113, Presented at the Canadian International Petroleum Conference (Annual Technical Meeting), Calgary, 7-9 June.

27. Mayerhofer, M.J., Ehlig-Economides, C.A., Economides, M.J., 1995. Pressure transient analysis of fracture calibration tests. J. Pet. Technol. 47 (3), 229-234.

28. Medeiros, F., Ozkan, E., Kazemi, H., 2008. Productivity and drainage area of fractured horizontal wells in tight gas reservoirs. SPE Reserv. Eval. Eng. 11 (5), 902-911. http:// dx.doi.org/10.2118/108110-PA. SPE-108110-PA.

29. Medeiros, F., Kurtoglu, B., Ozkan, E., Kazemi, H., 2010. Analysis of production data from hydraulically fractured wells in shale reservoirs. SPE Reserv. Eval. Eng. 13 (3), 559e568. http://dx.doi.org/10.2118/110848-PA. SPE-110848-PA.

30. Ouyang, L.B., Kikani, J., 2002. Improving Permanent Downhole Gauges (PDG) data processing via wavelet

analysis. In: Paper SPE 78290 presented at the European Petroleum Conference, Aberdeen, 29–31 October. http://dx.doi.org/10.2118/ 78290-MS.

31.   Ozkan, E., Raghavan, R., Apaydin, O.G., 2010. Modeling of fluid transfer from shale matrix to fracture network. In: Paper SPE 134830 Presented at the SPE Annual Technical Conference and Exhibition, Florence, Italy.

32.   Palacio, J.C., Blasingame, T.A., 1993. Decline curve analysis using type curvesanalysis of gas well production data. In: Paper SPE 25909 Presented at Joint Rocky Mountain Regional and Low Permeability Reservoirs Symposium, 26-28 April.

33.   Parmar, J.S., Dehghanpour, H., Kuru, E., 2012. Unstable displacement: a missing factor in fracturing fluid recovery. In: Paper SPE 162649 Presented at the SPE Canadian Unconventional Resources Conference, Calgary, Alberta, Canada, 30th October-1st November.

34.   Parmar, J.S., Dehghanpour, H., Kuru, E., 2013. Drainage against gravity: factors impacting the load recovery in fractures. In: Paper SPE 164530 Presented at the Unconventional Resources Conference, theWoodlands, Texas, USA, 10-12 April.

35.   Peres, A.M.M., Onur, M., Reynolds, A.C., 1993. A new general pressure analysis procedure for slug tests. SPE Form. Eval. 8 (4), 292e298 (December).

36.   Song, B., Ehlig-Economides, C., 2011. Rate-normalized pressure analysis for determination of shale Gas well performance. In: Paper SPE 144031 Presented at the SPE North American Unconventional Gas Conference and Exhibition, the Woodlands, Texas, USA, 14-16 June.

37.   Song, B., Economides, M.J., Ehlig-Economides, C., 2011. Design of multiple transverse fracture horizontal wells in shale gas reservoirs. In: Paper SPE 140555 Presented at the SPE Hydraulic Fracturing Technology Conference, the Woodlands, Texas, USA, 24-26 January.

38.   Williams-Kovacs, J.D., Clarkson, C.R., 2013. Stochastic modeling of two-phase flowback of multi-fractured horizontal

wells to estimate hydraulic fracture properties and forecast production. In: Paper SPE 164550 Presented at the SPE Unconventional Resources Conference, USA, the Woodlands, Texas, 10-12 April.

39. Zahid, S., Bhatti, A.A., Khan, H.H., Ahmad, T., 2007. Development of unconventional gas resources: stimulation perspective. In: Paper SPE 107053 Presented at the SPE Production and Operations Symposium, USA, Oklahoma City, Oklahoma, 31st March-3rd April.

Chapter 5

Modeling of Fluid Filtration and Near-wellbore Damage along a Horizontal Well

Supalak Parn-anurak and Thomas W. Engler

New Mexico Institute of Mining and Technology, United States

ABSTRACT

A method was developed to simulate drilling fluid invasion of a two-phase system (water-based mud in an oil-bearing formation) and to subsequently evaluate the damage along a horizontal well. The proposed model includes fluid invasion, filter cake buildup, and relative permeability components. Filter cake and mud filtration models were developed based on a mass balance equation of cake deposition and erosion and Darcy's equation. A convection–dispersion equation was solved numerically to characterize filtrate invasion. Exposure time, permeability anisotropy, and various formation properties are considered as factors causing non-uniform fluid invasion in horizontal wells. The results of the simulation provide the distribution of fluid filtrate and indicate the maximum invasion depths and the degree of fluid invasion around the wellbore. Coupling these results with relative permeability curves allows the assessment of the distribution of effective permeabilities. The effective permeability distribution and the depth of invasion are the keys to estimating the damage caused by fluid invasion. Damage radius and skin factor predicted by the model are in agreement with published results. Overall, this method provides improvement of fluid filtrate characterization and estimation of damage along a horizontal well.

INTRODUCTION

The use of horizontally drilled wells has increased dramatically during the past 10 years because they offer greater contact with the reservoir rocks. The effective production rate should be much greater for a horizontal well. Unfortunately, the larger drainage area and longer wellbore also contribute to a larger exposed area and longer exposure time for the drilling fluid in horizontal wells. Therefore, severe damage caused by fluid invasion can have a considerable influence on the reduction of horizontal well productivity (Renard and Dupuy, 1991). Consequently a better understanding of damage mechanisms, quantification of fluid invasion, and near-wellbore formation damage can minimize the risks of horizontal well drilling.

Prediction of fluid filtrate invasion and damage in horizontal wells is considered difficult and complicated because there are multiple factors that always exist in horizontal well drilling: these factors include variable contact times along the well and reservoir heterogeneity that can result in a non-uniform distribution of fluid filtrate and formation damage around horizontal wells. Thus, the quantification of the differing invasion radii along a horizontal well is very important.

A system of a water-based mud invading an oil-bearing formation is considered. A major damage mechanism of formation damage is the reduction of relative permeability surrounding the well due to changes of fluid saturation in the invaded zone. Based on the Hawkins (1956) equation, the reduction of permeability surrounding the wellbore and the invasion radius must be estimated to quantify the degree of damage.

Different models for estimation of near-wellbore damage in horizontal wells have been developed (Renard and Dupuy, 1991, Frick and Economides, 1993, Engler et al., 1995 and Yan et al., 1997). Most of the methods previously developed are limited to homogeneous reservoirs and are principally based on empirical data that result in less flexibility when they are applied under various conditions.

A numerical method for simulation of the filtrate invasion and prediction of the degree of damage surrounding a horizontal well is presented. An improved method proposed in this work is the coupling of wellbore and filtrate invasion models, which provide a new solution to simulate the fluid filtration and damage in horizontal wells under different wellbore conditions.

MODELING CONCEPT

Solid and fluid influx into the reservoir in an overbalanced pressure condition can cause permeability reduction from formation damage in the invaded zone. While the solid invasion is limited to a few inches away from the wellbore, the fluids will invade significantly deeper into the formation. Therefore, the invasion of fluids may be considered as the primary mechanism that causes the near-wellbore damage. Damage caused by fluid invasion includes phase trapping or blockage, wettability alteration, and emulsion blockage (Golan and Whitson, 1991).

Correct quantification of the damage is an important component for predicting well performance. While the damage in a vertical well is a function of permeability reduction in the damaged zone and damage radius, the magnitude of damage in a horizontal well depends on: 1) the filtrate distribution along a horizontal well, 2) the variation of permeability reduction surrounding the well, 3) the difference of overbalanced pressure as a result of the extended length of the well, and 4) longer contact time. To obtain an accurate assessment of the damage, fluid filtrate characterization and modeling along a horizontal well are required.

MODEL OVERVIEW

The solution of the filtration model is based on the assumption of steady-state radial flow (no flow from beneath the bit). Filtration rate reflects flow ability through the filter cake and formation. Build-up of a filter cake layer is explained by a mass balance equation of mud cake deposition and erosion. Combining Darcy's equation with the build-up model of the filter cake under a differential pressure condition will determine the fluid flow through filter cake and formation, as well as the filtration rate. The fluid distribution in the invaded zone is characterized by determination of the filtrate concentration using a convection–dispersion equation. With the filtration rate calculated, the convection–dispersion equation is solved numerically with given boundary conditions to characterize the movement of fluid filtrate into the formation with respect to time. The result shows the concentration profile of the fluid versus the invaded radius. The profile shows the distribution of fluid invasion and the radius of invasion, which are significant factors for assessing the degree of damage.

The displacement of oil by mud filtrate in the reservoir represents a two-phase system of fluids flowing though porous media. Consequently, the wetting phase saturations in the invaded zone are increased, resulting in the reduction of relative permeability of the oil phase. The displaced phase (oil) viscosity is assumed to be similar to the filtrate viscosity or close to 1 cP. For damage estimation, permeability–saturation curves for two-phase systems are used to couple the fluid saturation and relative permeability in the invaded zone. The assumption made here is that porosity/permeability is a constant. As for the mud filtration, they are reduced a lot. The degree of damage corresponds to the reduction of oil effective permeability due to the change of fluid saturation in the invaded zone.

FLUID FILTRATION MODEL

The model is based on the assumptions of single-phase radial flow of mud filtrate, constant rate of circulation, isothermal conditions, immiscible displacement, constant bottom-hole pressure, constant reservoir pressure, and no solid particle invasion, and, subsequently, unchanged porosity/permeability in invade zone. Since only the fluid component of drilling fluids is assumed to invade the formation, mud solid particles will deposit inside the wellbore. This deposition of the solid particles may generate a filter cake layer over the wellbore sand face Fig. 1. Therefore, the cake build-up is a strong function of the drilling fluid density. The mass of solid particles, m_c, deposited in the cake is determined by:

$$m_c = Ax_c(1 - \phi_c)\rho_c \tag{1}$$

During drilling, or circulation, mud filtration is in the dynamic period. The rate of cake build-up depends on the rate difference between cake deposition and cake erosion. An erosion rate is developed by the shear force of mud flowing along the cake surface. The cake deposition rate relies on the volume of fluid flux, which is a function of the filtration pressure gradient through the cake layer and formation permeability. When the filter cake gets thicker, the rate of cake build-up reduces until a stable condition between the rates of deposition and erosion is reached. Once stability occurs, the cake thickness becomes constant. The mass balance equation for the cake layer is explained as:

Rate of cake build-up

$$= \text{Rate of deposition } (d_c) - \text{Rate of erosion } (e_c) \tag{2}$$

The rate of deposition is proportional to the mass flux of solid particles in drilling fluids:

$$d_c = A u_{in} C_{solid} \tag{3}$$

The rate of erosion is related to the shear stress, τ, on the filter cake:

$$e_c = k_\tau A \tau \tag{4}$$

Consequently, the mass balance equation (Eq. (2)) can be written as Eq. (5) to define the build-up rate of filter cake:

$$\frac{dm_c}{dt} = A u_{in} C_{solid} - k_\tau A \tau \tag{5}$$

The cake build-up rate also can be defined as a differential of Eq. (1) with respect to time:

$$\frac{dm_c}{dt} = A \left(\frac{dx_c}{dt} \right) (1 - \phi_c) \rho_c \tag{6}$$

The mass rates of filter cake from Eqs. (5) and (6) are combined to define the cake build-up in terms of thickness change relating to volumetric flux:

$$\frac{dx_c}{dt} = \frac{u_{in} C_{solid} - k_\tau \tau}{(1 - \phi_c) \rho_c} \tag{7}$$

The initial volumetric filtration, u_{in}, at the wellbore sand face can be expressed as:

$$u_{in} = \frac{q_{in}}{2 \pi r_c h} \tag{8}$$

where r_c is the radius of the hole after the filter cake has been generated (distance between the center of the well to the filter cake surface). The cake thickness is a function of time and can be expressed as:

$$x_c(t) = r_w - r_c(t) \tag{9}$$

The initial filtrate flux, u_{in}, from Eq. (8) is substituted into Eq. (7) to yield:

$$\frac{\partial x_c}{\partial t} = aq(t) + b \tag{10}$$

where:

$$a = -\frac{1}{2}\frac{C_{\text{solid}}}{\rho_c(-1 + \phi_c)\pi r_c h} \tag{11}$$

$$b = \frac{k_\tau \tau}{\rho_c(-1 + \phi_c)} \tag{12}$$

The generation rate of filter cake depends on filtration rate, q. In order to solve for the cake thickness, the filtration rate (which is a function of time) must be defined. Fig. 1 shows a simple model of the filter cake in a radial flow system.

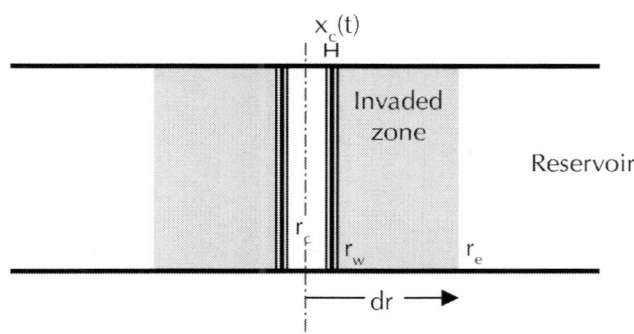

Figure 1: An illustration of filter cake and damage zone in a radial flow system.

Applying Darcy's equation, the pressure gradient equation for a radial flow system is:

$$\frac{\partial p}{\partial r} = -\frac{u_r \mu}{k} \tag{13}$$

Integration of Darcy's equation from $p_w...p_e$, $r_w...r_e$ at t=0, x_c=0 results in:

$$p_w - p_e = \frac{1}{2} \frac{q_0 \mu \ln\left(\frac{r_e}{r_w}\right)}{\pi h k} \tag{14}$$

where q_0 is the initial filtration rate.

Integration of Darcy's equation from $p_{w(x_c>0)} \dots p_e$, $r_{w(x_c>0)} \dots r_e$ at $t=t$, $x_c=x_c(t)$ results in:

$$p_{w(x_c>0)} - p_e$$

$$= \frac{1}{2} \frac{q(t)\mu\left(\ln\left(\frac{r_e}{r_w}\right) + \frac{k\ln\left(\frac{r_w}{r_w - x_c(t)}\right)}{k_c}\right)}{\pi h k} \tag{15}$$

where $p_{w(x_c>0)}$ is the pressure at the mud cake face inside the wellbore.

Because the wellbore and reservoir pressures are assumed to be constant during the build-up of the filter cake, Eqs. (14) and (15) are equivalent at any time. Thus, the fluid invasion rate, q, can be solved by the combination of Eqs. (14) and (15):

$$q(t) = \frac{q_0 \ln\left(\frac{r_e}{r_w}\right) k_c}{\ln\left(\frac{r_e}{r_w}\right) k_c + k\ln\left(\frac{r_w}{r_w - x_c(t)}\right)} \tag{16}$$

The filtration rate change with time from Eq. (16) is substituted into Eq. (10). The equation for cake development with time becomes:

$$\frac{\partial x_c}{\partial t} = \frac{\frac{1}{2} \frac{q_0 \ln\left(\frac{r_e}{r_w}\right) k_c C_{solid}}{\left(\ln\left(\frac{r_e}{r_w}\right) k_c + k\ln\left(\frac{r_w}{r_w - x_c(t)}\right)\right)\pi r_c h} - k_\tau \tau}{(1 - \phi_c)\rho_c} \tag{17}$$

Eq. (17) can be discretized into:

$$x_c(t + \Delta t) - x_c(t)$$

$$= \frac{\left(\frac{1}{2}\frac{q_0 \ln\left(\frac{r_e}{r_w}\right)k_c C_{solid}}{\left(\ln\left(\frac{r_e}{r_w}\right)k_c + k\ln\left(\frac{r_w}{r_w - x_c(t)}\right)\right)\pi r_c h} - k_\tau \tau\right)\Delta t}{(1 - \phi_c)\rho_c}$$

(18)

where t=0,1,2,..., (TMAX− t). TMAX is maximum exposure time for a given interval, synonymous with reaching an equilibrium state.

Eq. (18) is used to calculate the cake thickness for the next time step. For example, if the time step size equals 1, x_c at t=1 can be calculated while q_0 and x_c at t=0 are known.

Filtration rate at the next time step can be solved explicitly after the filter cake thickness is estimated. Eq.(16) of the next time step can be expressed as:

$$q(t + \Delta t) = \frac{q_0 \ln\left(\frac{r_e}{r_w}\right)k_c}{\ln\left(\frac{r_e}{r_w}\right)k_c + k\ln\left(\frac{r_w}{r_w - x_c(t + \Delta t)}\right)}$$

(19)

where t=0,1,2,..., (TMAX− t).

FILTRATE DISTRIBUTION MODEL

To characterize fluid filtrate in a formation, filtrate distribution in the invaded zone must be considered. The filtrate concentration can vary radially from the wellbore depending on the dispersitivity of formation, formation porosity, fluid velocity flux, and time. Assuming unchanged formation porosity and incompressible fluid, Civan and Engler (1994) expressed the relationships among these variables as:

$$\frac{1}{r}\frac{\partial}{\partial r}\left(rD\frac{\partial C}{\partial r}\right) - \frac{u}{\phi(1-S_{or})}\frac{\partial C}{\partial r} = \frac{\partial C}{\partial t}$$

(20)

The initial conditions are:

$C(r_w,0)=C_f$

$C(r,0)=0$

The boundary conditions, at t>0 and q>0, are written as:

$C(r_w,t) = C_f$ (inner boundary condition)

$C(r_e,t) = 0$ (outer boundary condition)

The fluid velocity flux, u, in the concentration equation (Eq. (20)) is defined as a filtration rate over a fluid-exposed area at the wellbore:

$$u(t) = \frac{q(t)}{2\pi r_i h}$$

(21)

where q can be solved by Eq. (19) with respect to time and i=1,2,..., r_e-r_w (grid size equals one).

The dispersion coefficient, D, consists of two components: convection dispersion, D_e, and molecular diffusion, D_m, and was expressed by Donaldson and Chernoglazov (1987) as:

$$D = D_e + D_m$$

(22)

The molecular diffusion term is considered negligible since the convection term is the significant term during fluid filtration. Several authors have established an exponential relationship between D and the fluid velocity (u.) as (Perkins and Johnson, 1963 and Baker, 1997):

$$D = f u^g$$

(23)

The variables f and g are empirical parameters. The average values of f and g determined in Poulin's (1985) work are 1.25 and 51.7, respectively.

An implicit formulation is applied for the numerical solution. By applying inner and outer boundaries, Eq(20) can be solved using finite difference (Appendix A). The solution provides the filtrate concentration located at each grid surrounding the wellbore.

REDUCTION OF PERMEABILITY IN THE DAMAGE ZONE

The filtrate concentration distribution within the investigated radius for an exposure time is characterized by solving a time and space matrix. Because only water-based mud is considered in this study, filtrate concentration reflects the magnitude of water saturation. The fluid filtrate invasion increases water saturation in the invaded zone. This circumstance contributes to the phase trapping effect (i.e., an increase in the trapped fluid saturation in the porous media around the wellbore). The consequence is a substantial decrease of relative permeability to the displaced phase. This problem has been recognized to cause severe loss of well productivity and be a main reason for wellbore damage (Bennion, 1999).

The change of water saturation is a function of the filtrate concentration in the formation, which is related to residual oil saturation and irreducible water saturation through Eq. (24):

$$S_{\text{mud}}(r) = C_f(r)\left(S_{w,\max} - S_{w,\min}\right) + S_{w,\min}$$

$$(24)$$

The degree of permeability reduction reflects how much is the decrease of relative permeability of the hydrocarbon phase in the invaded zone.

The power law functions modified after Brooks and Corey (1994) are applied to describe the relationship of filtrate saturation and relative permeability:

For oil: $k_{ro} = k_{ro,max} \left(\dfrac{S_o - S_{or}}{1 - S_{wi} - S_{or}} \right)^{no}$

$\quad\quad = k_{ro,max} \left(\dfrac{1 - S_{mud} - S_{or}}{1 - S_{wi} - S_{or}} \right)^{no}$

$$\tag{25}$$

For water: $k_{rw} = k_{rw,max} \left(\dfrac{S_{mud} - S_{wi}}{1 - S_{wi} - S_{or}} \right)^{nw}$

$\quad\quad = k_{rw,max} \left(\dfrac{1 - S_{wi} - S_o}{1 - S_{wi} - S_{or}} \right)^{nw}$

$$\tag{26}$$

The effect of saturation changes of the relative permeability results in the formation of damage due to phase trapping or blockage. Therefore permeability in the damage zone varies with the filtrate saturation. The equation used to estimate the damage can be written as (derivation in Appendix B):

$$s = k_{eff} \int_{r_w}^{r_d} \left(\frac{1}{r k_{d,eff}(r)} \right) dr - \ln\left(\frac{r_d}{r_w} \right) \tag{27}$$

The distribution of permeability reduction corresponds to the decrease of relative permeability in the invaded zone, and the invasion depth (shown in Fig. 2). The permeability reduction is solved numerically by coupling the filtrate distribution model with the relative permeability correlation. A mean value of the various permeability zones, $\overline{K}_d(r)$, can be determined by numerical integration:

$$\overline{k}_d = \frac{(r_{d,1} - r_w)k_{d,1} + (r_{d,2} - r_{d,1})k_{d,2} + \cdots + (r_d - r_{d,n})k_{d,n}}{(r_d - r_w)} \tag{28}$$

Thus, the skin factor is:

$$s = \frac{k_{\text{eff}}}{k_{d,\text{eff}}} \ln\left(\frac{r_d}{r_w}\right) - \ln\left(\frac{r_d}{r_w}\right)$$

(29)

which represents the damage that results from the effective permeability reduction in the invaded zone. Generally, skin due to high velocity flow or non-Darcy flow can also be an important part of total skin if production rates are high or the open-flow interval is only a small portion of the formation thickness. Compared to a vertical well, however, skin damage can be the most severe in an openhole horizontal well because of the much longer exposure time to mud fluids. Meanwhile, the damage induced by high velocity flow can be less critical for a horizontal well due to a larger contact area and a lower drawdown pressure.

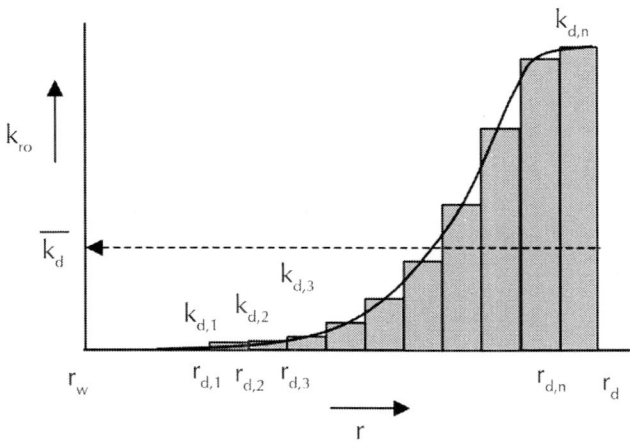

Figure 2: Numerical approximation of the average permeability in the damage zone.

Each different section of a horizontal well should have a different magnitude and radius of damage because the fluid invasion in a horizontal well is a function of exposure time. Considering each section of the well as an investigated node, the model can be used to evaluate the damage distribution along a horizontal well. As shown in Fig. 3, each node along the horizontal well represents the damage distribution near wellbore at a different exposure time.

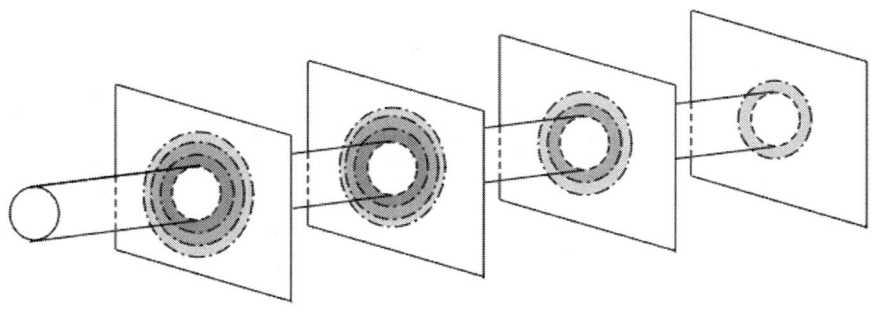

Figure 3: A schematic of filtrate saturation at each node surrounding a horizontal wellbore.

The exposure time at each node depends on rate of penetration and invasion time only. These conditions are based on the assumptions that invasion occurs in a homogeneous rock and that the wellbore exhibits infinite conductivity behavior. If these assumptions do not hold, the reservoir properties such as permeability and porosity in the model and the wellbore pressure at each node will not be constant, which results in a significant pressure loss in the wellbore.

The filtration rate at the sandface in an anisotropic medium can be estimated by applying the proposed model because the initial invasion rate is considered at the wellbore (which is similar in both isotropic and anisotropic cases). Data such as permeability in horizontal and vertical directions, and overbalanced pressure are required. Because isotropy and anisotropy are relevant only to the formation permeability in this study, the area of the fluid invasion will be parameterized from circular radial-flow domain to axially symmetric radial-flow domain, which would have major and minor axes depending on the magnitude of permeability.

APPLICATION AND DISCUSSION OF RESULTS

To verify the proposed model, two comparisons were performed using the published data from Yan et al. (1997). In the first test, invasion depths were measured from 30 min of filtration in 13 core samples and the results were compared to Yan et al.'s empirical method and to the proposed model. The second test compares the estimated invasion depth and skin in a horizontal well between the reference, empirical method, and numerical model. In both cases, static filtration conditions exists (i.e., no mud circulation occurs in the wellbore). Pressure differential, permeability, and porosity were measured on 13 core samples. Porosity ranged from 7.5% to 29.5%, permeability from 33 to 710 md, and pressure differential from 221 to 573 psi, respectively. These laboratory-measured parameters were obtained to establish a regression fit to develop an empirical method (Yan et al.) for estimating invasion depth. Also obtained were invasion depths after 30 min of filtration for comparison to the empirical method. Fig. 4 illustrates the depth of invasion from the experimental measurements, the empirical method, and the proposed numerical model, and the percent error between the measured and numerical model results.

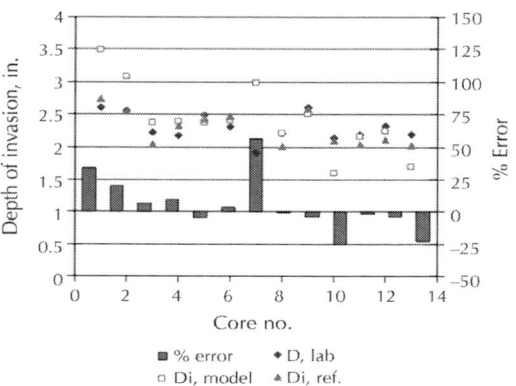

Figure 4: Comparison of experimental, empirical, and numerical invasion depths under static conditions from 13 core samples.

As expected, the empirical method provides a good fit to the measured data, with an invasion depth range of 1.9–2.7 in. for both. However, no blind test was attempted; therefore, the empirical method is limited to the data range and set that developed the correlation.

The results from the numerical model are reasonable, with a range of 1.7–3.5 in. for the various core samples. Notice in Fig. 4 that in five samples, the model results have significant error, while the invasion depth results in the remaining eight samples were in agreement with the measurements. A parametric study revealed that the magnitude of the error is directly related to the formation permeability. That is, the highest positive error (samples 1, 2, and 7 in Fig. 4) is associated with the highest permeability samples (>379 md) and the lowest negative error (samples 10 and 13) is associated with the lowest permeability samples (<79 md). It is speculated that the over-estimation of invasion depths in high permeability formations and the under-estimation in low permeability formations is the results of inaccurate input of filter cake properties. Filter cake permeability, porosity, and density were held constant for all samples. These properties should be adjusted to account for the higher and lower formation permeabilities. Additional details on the parametric analysis and discussion can be found in Parnanurak (2003). In the second test, an estimation of the damage in a horizontal well cited in the reference paper is compared with the damage estimated from the numerical model. The radius of damage at each segment of the well, the damage permeability, and the skin factor of the investigated horizontal well from the Dagang oilfield in east China were calculated. The horizontal length of the well is 207 m long with 120 h of total drilling time. Other parameters and reservoir characteristics are shown in Table 1.

Table 1: Given parameters and reservoir data of a horizontal well (Yan et al., 1997)

Horizontal length, HL	207 m
Wellbore radius, r_w	0.22 m

Maximum deviation angle, ϑ_{max}	86°
Reservoir thickness, h	9 m
Horizontal permeability, K_h	0.1 μm^2
Vertical permeability, K_v	0.035 μm^2
Formation porosity, $\varnothing$	0.23
Overbalanced drilling pressure, ΔP	3 MPa

Permeabilities from the damaged and undamaged core samples for the horizontal well were measured (Yan et al., 1997 and Yan et al., 1998). The filtration is a two-phase flow process (oil saturated cores displaced by water-based mud); therefore the measured permeability represents the oil effective permeability. The ratio of damaged to undamaged permeability for the core samples is 81.3%, which means that about 20% oil effective permeability is reduced by the filtration. The radius of damage predicted by the empirical method ranged from 0 to 40.2 cm at different points along the horizontal section. The permeability reduction, coupled with the estimated damage radius, was used to calculate the skin factor at each point of the horizontal well by applying the Hawkins equation (Eq. (B4), Appendix B). The mean skin factor of the well is estimated to be about 0.19.

With the given input data, the proposed model was applied to simulate the invasion of fluid into a horizontal well and estimates the radius of damage. Because the empirical correlation applied is derived under the condition of static filtration, the erosion term (Eq. (4)) in the proposed model does not exist in this case. The additional data used in the model are shown in Table 2.

Table 2: Additional data needed for the damage estimation

Cake permeability, k_c	1×10^{-9} μm^2
Cake porosity, $\varnothing_c$	0.1
Cake density, ρ_c	2440 kg/m³
Mud viscosity, μ	0.02 Pa s
Mud concentration, C_{mud}	1090 kg/m³

Filtrate concentration, C_f	1000 kg/m³
Irreducible water saturation, S_{wi}	0.25
Residual oil saturation, S_{or}	0.3
Oil relative permeability, $K_{ro,max}$	1
Oil exponent in the power-law expression, no	2

Shown in Table 3 and Fig. 5 is the calculated damage radius from the numerical model compared to the published results. The results show that the estimated invasion radius ranges from 0 to 42.4 cm. The maximum invasion radius (42.4 cm) is at the heel of the horizontal well, with no invasion at the toe. The damage radius estimated by the proposed model is approximately 5% smaller at the toe and 6% larger at the heel than the empirical method. Oil effective permeability in the damage zone ranges from 72% to 81% of the original permeability (from heel to toe). The mean permeability is 76.3% of the original permeability and the mean skin equals 0.18. Comparing damage permeability and skin to those with the reference data (81.3% for the damage permeability and 0.19 for the skin factor) confirms the validity of the proposed model.

Both methods represent the filtration of a two-phase process (oil-bearing formation replaced by water-based mud); therefore damage caused by reduction of relative permeability is dominant. Based on the assumptions that phase trapping is the major cause of damage and no chemical reaction occurs in the filtration process, the degree of damage is mainly dependent on the fluid relative permeability effect. The different filtrate properties result in different relative permeabilities of water to oil phases and can be a significant factor affecting the degree of damage. Adjusting the relative permeability curve in the proposed model reflects the change of oil relative permeability for the filtrate properties.

Table 3: Damage radius along the examined horizontal well

L (m)	0	20	40	60	80	100	120	140	160	180	207
Time (h)	120	104.1	92.5	81	69.4	57.8	46.3	34.7	23.1	11.6	0
Time (min)	7200	6246	5550	4860	4164	3468	2778	2082	1386	696	0
$R_{s,ref}$ (cm)	40.2	38.9	38.6	38.3	38.0	37.6	37.1	36.5	37.0	34.4	0
$R_{s,mod}$ (cm)	42.4	41.9	40.8	40.1	39.6	37.9	37.5	35.6	35.0	32.9	0
% Difference	5.6	7.7	5.7	4.7	4.1	0.8	1.2	−2.5	−5.5	−4.2	0

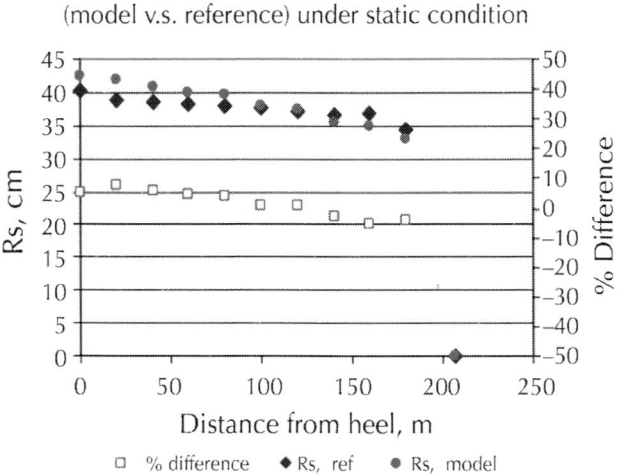

Figure 5: A comparison of damage radii from the reference and the proposed model.

CONCLUSIONS

The model that was developed can be used to simulate the radial flow of water-based filtrate through oil-bearing formations during static or dynamic filtration. The assumptions of no solid particle invasion into the formation along with no chemical reaction terms simplify the model. Filter cake and mud filtration models are developed based on a mass balance equation of cake deposition and erosion. A convection–dispersion equation is solved numerically to characterize the filtrate invasion behavior. An analogous Hawkins equation is applied to predict the degree of damage in a horizontal well.

A range of invasion radii exists along a horizontal well due to the different contact times at each part of the well. The degree of damage is reflected by the invasion radius and the reduction of relative permeability resulting from the change of fluid saturation surrounding the wellbore. The model testing is performed by predicting the damage radius and skin factor in a horizontal well.

The predictions of damage radii and skin factor are compared and show a good match with published results.

With independent input of formation properties at each node along the well, the model can be used to predict the fluid invasion and variable damage along the well in a heterogeneous reservoir.

Appendix A Numerical Solution for the Concentration Equation

$$\frac{\partial C}{\partial t} = \frac{1}{r}\frac{\partial}{\partial r}\left(rD\frac{\partial C}{\partial r}\right) - \frac{u}{\phi(1-S_{or})}\frac{\partial C}{\partial r} \tag{A1}$$

can be written as:

$$\frac{\partial C}{\partial t} = \frac{1}{r}\left(\left(r\frac{\partial D}{\partial r}+D\frac{\partial r}{\partial r}\right)\frac{\partial C}{\partial r}+rD\frac{\partial^2 C}{\partial r^2}\right)$$

$$-\frac{u}{\phi(1-S_{or})}\frac{\partial C}{\partial r} \tag{A2}$$

where $u=\dfrac{q(t)}{2\pi r h}$ and $D=fu^g$. Introducing Eqs. (21) and (23) to Eq. (A2) results in:

$$\frac{\partial C}{\partial t} = \left(\frac{1}{r}\left(f(1-g)\left(\frac{q(t)}{2\pi r h}\right)^g\right)-\frac{\left(\dfrac{q(t)}{2\pi r h}\right)}{\phi(1-S_{or})}\right)\frac{\partial C}{\partial r}$$

$$+f\left(\frac{q(t)}{2\pi r h}\right)^g\frac{\partial^2 C}{\partial r^2} \tag{A3}$$

Eq. (A1) becomes:

$$\frac{\partial C}{\partial t} = \alpha\frac{\partial C}{\partial r}+\delta\frac{\partial^2 C}{\partial r^2} \tag{A4}$$

where:

$$\alpha = \frac{1}{r}\left(f(1-g)\left(\frac{q(t)}{2\pi rh}\right)^g\right) - \frac{\left(\dfrac{q(t)}{2\pi rh}\right)}{\phi(1-S_{or})} \tag{A5}$$

$$\delta = f\left(\frac{q(t)}{2\pi rh}\right)^g \tag{A6}$$

By applying linear finite difference method, the difference formula for the concentration equation, Eq.(A4), is stated as:

$$\frac{\partial^2 C_i}{\partial r^2} = \frac{C_{i+1}^{t+1} - 2C_i^{t+1} + C_{i-1}^{t+1}}{(\Delta r)^2} \tag{A7}$$

Applying implicit formulation to solve for the numerical solution:

$$\left(\frac{\delta_i^{t+1}\Delta t}{(\Delta r)^2} - \frac{\alpha_i^{t+1}\Delta t}{2\Delta r}\right)C_{i-1}^{t+1} - \left(2\frac{\delta_i^{t+1}\Delta t}{(\Delta r)^2} + 1\right)C_i^{t+1}$$

$$+\left(\frac{\delta_i^{t+1}\Delta t}{(\Delta r)^2} + \frac{\alpha_i^{t+1}\Delta t}{2\Delta r}\right)C_{i+1}^{t+1} = -C_i^t \tag{A8}$$

where $\eta = \dfrac{\Delta t}{\Delta r^2}$ and $\xi = \dfrac{\Delta t}{2\Delta r}$ and denoting:

$$A_i' = \left(\delta_i^{t+1}\eta - \alpha_i^{t+1}\zeta\right), \quad B_i' = -\left(2\delta_i^{t+1}\eta + 1\right),$$

$$C_i' = \left(\delta_i^{t+1}\eta + \alpha_i^{t+1}\zeta\right), \quad A_i'C_{i-1}^{t+1} + B_i'C_i^{t+1}$$

$$+ C_i'C_{i+1}^{t+1} = -C_i^t \tag{A9}$$

where i=1,2,...,N and t=1,2,...,TMAX.

The initial condition at t= 0 gives:

$$C_i = C_f, \qquad i = 0$$

$$C_i = 0, \qquad i = 1, 2, ..., N$$

The inner and outer boundary conditions at t=1,2,...,TMAX and q>0 can be stated as:

$$C_i = C_f, \qquad i = 0$$

$$C_i = 0, \qquad i = r_e$$

At i=1 and t=1,2,...,TMAX, Eq. (A9) becomes:

$$B_1' C_1^{t+1} + C_i' C_2^{t+1} = -C_1^t - A_1' C_f$$

where $C_0 \cong C_f$ at any time (Fig. A1).

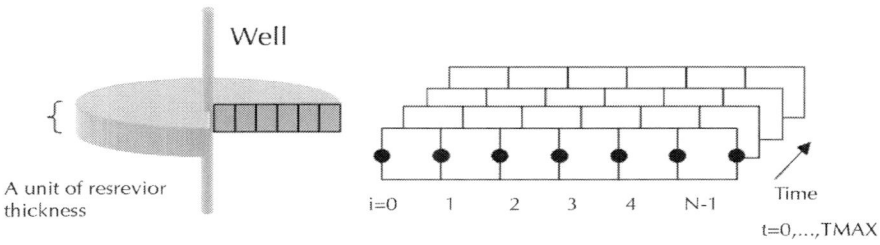

Figure A1: Time and space domain and the point distributed grid system.

The outer boundary condition gives:

$$C_N = 0 \quad \text{and} \quad C_{N+1} = 0$$

where N=r$_e$ (maximum invasion radius).

Therefore, at the outer boundary, i=N and t=1,2,...,TMAX, Eq. (A9) becomes:

$$A_N' C_{N-1}^{t+1} + B_N' C_N^{t+1} = -C_N^t - C_N' C_{N+1}^{t+1}$$

$$A_N' C_{N-1}^{t+1} + B_N'(0) = 0$$

For i=1,2,3,...,N−1, N, a form of linear matrix equations, can be written as:

$$
\begin{bmatrix}
B'_1 & C'_1 & 0 & A & A & 0 \\
A'_2 & B'_2 & C'_2 & 0 & A & 0 \\
0 & A'_2 & B'_2 & C'_2 & 0 & 0 \\
M & O & O & O & O & M \\
0 & A & 0 & A'_{N-1} & B'_{N-1} & C'_{N-1} \\
0 & A & A & 0 & A'_N & B'_N
\end{bmatrix}
\begin{bmatrix}
C^{t+1}_1 \\
C^{t+1}_2 \\
C^{t+1}_3 \\
M \\
C^{t+1}_{N-1} \\
C^{t+1}_N
\end{bmatrix}
$$

$$
= -
\begin{bmatrix}
C^t_1 + A'_1 C_{in} \\
C^t_2 \\
C^t_3 \\
M \\
C^t_{N-1} \\
0
\end{bmatrix}
$$

Note that the right-hand side of the matrix will appear in the computer program as Di where i=1,2,3,...,N.

Appendix B. Derivation of Skin Factor

This additional section describes the concept of skin effect relating to the distribution of permeability reduction around the wellbore. Eq. (B1) expresses a differential form of Darcy's equation for a steady-state cylindrical flow, uniform permeability in all direction, at an arbitrary distance, r:

$$
v = \frac{qB}{2\pi rh} = \frac{k}{\mu}\frac{dp}{dr}
\tag{B1}
$$

Fig. B1 illustrates a radial pressure distribution for an ideal well (dash line) and an actual well (solid line).

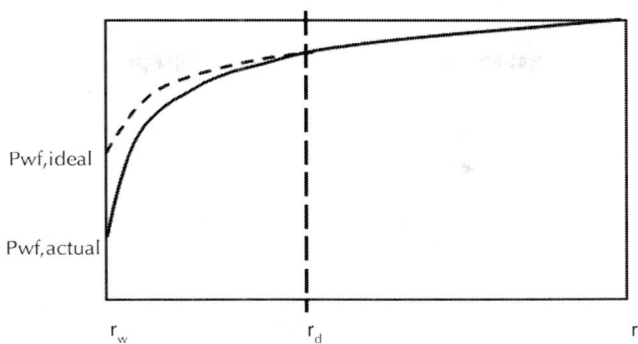

Figure B1: Pressure distribution profiles of a well with and without near-wellbore damage.

Hawkins stated that if a uniform permeability, k_d, in a damage zone, r_d, is presented, the additional pressure drop across the zone, p_s, can be described by the steady-state radial flow equation. Thus:

$$dp_{skin} = dp_{actual} - dp_{ideal} = \frac{qB}{2\pi hr}\frac{\mu}{k_d}dr - \frac{qB}{2\pi hr}\frac{\mu}{k}dr \qquad (B2)$$

Or

$$dp_{skin} = \frac{q\mu B}{2\pi kh}S \qquad (B3)$$

where $\quad S = \frac{1}{r}(\frac{k}{k_d} - 1)dr$.

In case a uniform extent of the damage (constant damage permeability) is presented from r_w to r_d, the skin factor can be expressed as:

$$s = \left(\frac{k}{k_d} - 1\right)\ln\left(\frac{r_d}{r_w}\right) \qquad (B4)$$

In case a variable damage zone (damage permeability as a function of radius) is presented from r_w to r_d, the skin factor is:

$$s = k \int_{r_\mathrm{w}}^{r_\mathrm{d}} \left(\frac{1}{r k_\mathrm{d}(r)} \right) \mathrm{d}r - \ln\left(\frac{r_\mathrm{d}}{r_\mathrm{w}} \right)$$

(B5)

Notice that from the skin equations, permeability reduction ratio has more influence to the damage than to the depth of damage as the damage radius is under logarithmic term.

REFERENCES

1.   Baker, L.E., 1997. Effects of dispersion and dead-end pore volume in miscible flooding. Soc. Pet. Eng. J. 17 (3), 219 – 227.

2.   Bennion, B., 1999. Formation damage—the impairment of the invisible, by the inevitable and uncontrollable, resulting in an indeterminate reduction of the unquantifiable. J. Can. Pet. Technol. 38 (2), 11 – 15 (Feb.).

3.   Brooks, R.H., Corey, A.T., 1994. Hydraulic properties of porous media. Hydrology Paper, vol. 3. Colorado State University, Ft. Collins, Co.

4.   Civan, F., Engler, W.T., 1994. Drilling mud filtrate invasion— improved model and solution. J. Pet. Sci. Eng. 11, 183 – 193.

5.   Donaldson, E.C., Chernoglazov, V., 1987. Characterization of drilling mud fluid invasion. J. Pet. Sci. Eng. 1, 3 – 13.

6.   Engler, W.T., Osisanya, S., Tiab, D., 1995. Measuring skin while drilling. SPE 29526, Production Operations Symposium, Oklahoma City, OK, April 2–4.

7.   Frick, T.P., Economides, M.J., 1993. Horizontal well damage characterization and removal. SPE Prod. Facil. 15 – 22 (Feb.).

8.   Golan, M., Whitson, H.C., 1991. Well Performance. Prentice-Hall, Inc., Englewood Cliffs, New Jersey, pp. 241.

9.   Hawkins, M.F., 1956. A note on the skin effect. Trans. AIME 207, 356 – 357.

10.   Parn-anurak, S., 2003. Modeling of fluid filtration and

nearwellbore damage along a horizontal well. PhD dissertation, New Mexico Institute of Mining and Technology, Socorro, NM, December.

11.   Perkins, T.K., Johnson, O.C., 1963. A review of diffusion and dispersion in porous media. Soc. Pet. Eng. J., 70 – 84 (March).

12.   Poulin, T., 1985. The determination of the coefficient of dispersion in sandstones. MS thesis, University of Oklahoma, Norman, OK, p. 112.

13.   Renard, G., Dupuy, J.G., 1991. Formation damage effects on horizontal-well flow efficiency. J. Pet. Technol., 786 – 869 (July).

14.   Yan, J., Jiang, G., Wu, X., 1997. Evaluation of formation damage caused by drilling and completion fluids in horizontal wells. J. Can. Pet. Technol. 36 (5), 36 (May).

15.   Yan, J., Jiang, G., Wang, F., Fan, W., Su, C., 1998. Characterization and prevention of formation damage during horizontal drilling. SPE Drill. Complet. 13 (4), 243 – 249.

6

Prediction of Drop Size Distribution in a Horizontal Pulsed Plate Extraction Column

M. Khajenoori[a, b], A. Haghighi-Asl[a], J. Safdari[b], and M.H. Mallah[b]

[a]School of Chemical, Petroleum and Gas Engineering, Semnan University, Semnan, Iran
[b]Nuclear Fuel Cycle Research School, Nuclear Science and Technology Research Institute, Tehran, Iran

ABSTRACT

Mean drop size and drop size distribution in a horizontal pulsed plate extraction column were investigated using different four binary systems. The effects of pulse intensity (af) and flow rates of both liquid phases have been investigated. The drop size decreased more rapidly with the increase of pulse intensities. It was observed that an increase in intensity of the pulses will lead to narrower ranges of distribution for the drop size. Increasing the flow rate of dispersed phase tends to increase the drop size. The effect of continuous phase flow rate is weaker than the effect of the dispersed phase flow rate. By using results, a semi empirical correlation obtained for the estimation of mean drop size which proves to be in good agreement to the experimental data. The average absolute relative error (AARE) of this correlation is about 15.6%. In order to find a predictive correlation for drop size distribution, four models of distribution functions are tested. The normal probability density function is the only suitable way for representing the experimental drop size distributions with an AARE of 13.7%.

INTRODUCTION

Liquid–liquid extraction is one of the classical methods in separation technology and finds applications in the chemical and petroleum industry, hydrometallurgy, biotechnology, nuclear technology, food industry, waste management, and other areas [1], [2], [3], [4], [5] and [6]. The efficiency of liquid–liquid contactors is primarily dependent on the degree of turbulence imparted to the system and the interfacial area available for mass transfer. The rate of mass transfer can be enhanced by pulsating motion imparted to the liquids by an external mechanical or electronic device. Internal mechanical parts are eliminated, leakage is minimized, and the pulsator can be isolated [7]. The pulsed columns have a clear advantage over other mechanical contractors when processing corrosive or radioactive solutions, since the pulsing unit can be remote from the column. The absence of moving mechanical parts in such columns slightly obviates repair and servicing. These advantages have led to the application of these columns in chemical, biochemical and petroleum industries [8], [9] and [10]. The pulsed liquid–liquid extraction columns can be divided as follows:

- The vertical pulsed columns
    a. The pulsed sieve plate columns.
    b. The pulsed packed columns.
    c. The pulsed disc and doughnut columns.

    d. The pulsed spray columns.

- The horizontal pulsed columns

    a. The horizontal sieve plate columns.

    b. The horizontal packed columns.

The main advantages and disadvantages of the two types of columns in identical conditions are as follows:

- Throughput of the horizontal pulsed columns is less than the vertical pulsed columns [11].
- The requirement time to start up horizontal pulsed columns is more than the vertical pulsed columns.
- The vertical pulsed columns are proper for industries that there is surface area limitation [12],[13] and [14].
- The horizontal pulsed columns are proper for industries that there is height limitation [15].
- The mass transfer efficiency for both types of columns is comparable [16].

Therefore, the advantages of vertical pulsed column are more than the horizontal pulsed columns but when we cannot use vertical pulsed column because of height limitation for installation as indoor, the horizontal pulsed column can be a proper contactor.

It is well known that the mean drop size and drop size distribution are important parameters in the study and design of the extraction column [17]. In the modeling of liquid–liquid extraction columns, where a dispersed phase exists as discrete drops, an average volume surface diameter is commonly used to predict the interfacial area for mass transfer, contact times and mass transfer coefficients [18], [19], [20] and [21]. In order to develop appropriate design procedures for a given type of extraction column, a knowledge of average drop size in terms of operating parameters, liquid physical properties is thus of paramount importance. Some investigations have been carried out on the drop size and drop size distribution in vertical pulsed column. Yadav and Patwardhan [22] presented a review on the drop size. Gholam Samani et al. [23] presented an experimental study of the Sauter mean drop size in a pulsed packed extraction column. Usman et al. [24] investigated the effect of the pulse intensity, and the dispersed phase and continuous phase velocities on the Sauter mean diameter. However, it is rare to find such reports about the drop size and drop size distribution of the HPC. A limited investigation has been carried out on the holdup and throughput capacity of the HPC.

It is the first time that, the effect of operating parameters in horizontal pulse column has been studied on mean drop size and drop size distribution. By the use of resulting data two empirical correlations have been proposed for the Sauter mean drop diameter and drop size distribution as a function of operating parameters, physical properties of the liquid systems and active column height.

The average absolute relative error (AARE) was used as an objective function to calculate the fitted parameters:

$$AARE = \frac{1}{n} \sum_{i=1}^{n} \frac{|X_i(exp) - X_i(thoe)|}{X_i(exp)}$$

(1)

where n is the number of data points, and $X_i(exp)$ and $X_i(theo)$ represent the experimental and theoretical data, respectively.

MATERIAL AND METHODS

Description of Equipment

Experiments were carried out in a semi-industrial horizontal pulsed sieve plate column. The material used for construction was glass. The active part of column was a pipe housing an internal plate cartridge consisting of 25 pairs of sieve plates constructed from 304 stainless steel. Each plate has perforations 2 mm in diameter, located only over half of the available plate area with a net free area of 22%. They contained 106 circular holes laid on triangular pitch of 4 mm.

The sieve plates were arranged as pairs with 1.0 cm spacing between each plate. The perforations on the plate nearest the light-phase inlet were located on the bottom and the perforations on the plate nearest the heavy-phase inlet were located on the top. Individual cells inside the column were created by spacing the plate pairs 5.0 cm apart. The column characteristics are listed in Table 1. A settler of 9 cm diameter at both ends of the column was employed to separate the two liquid phases. The inlet and outlet streams of the column were connected to four tanks, each of 25 l capacity. The flow rates of the two phases were measured by two rotameters.

Table 1: Geometrical characteristics of the column used

Material used for settler and column	Glass
Column length (m)	1.46
Column diameter (cm)	6.2
Upper and lower settler diameter (cm)	9
Upper settler length (cm)	60
Lower settler length (cm)	30

Material used for plate	Stainless steel
Plate thickness (mm)	1
Hole diameter (mm)	2
Hole pitch (mm)	4
Spacing the plate pairs (cm)	5
Spacing between each plate (cm)	1
Fractional free area (–)	0.22

The pulsator is an air pulsing system. The interface location of two phases at the top of the heavy-phase inlet and under light-phase outlet in upper settler was automatically controlled by an optical sensor. A solenoid valve (a normally closed type) was provided at the outlet stream of heavy phase. This valve received electronic signals from the optical sensor. When the interface location was going to change, the optical sensor sent a signal to solenoid valve and the heavy phase was allowed to leave the column by opening the diaphragm of solenoid valve. The light phase was allowed to leave the column via an overflow. A schematic diagram of experimental apparatus is shown in Fig. 1.

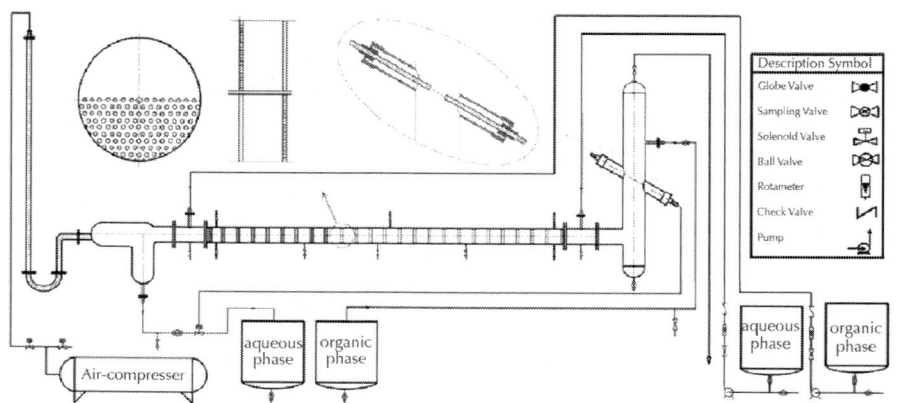

Figure 1: A schematic diagram of the horizontal pulsed sieve plate column.

Liquid–liquid Systems Used

Four liquid–liquid systems were chosen to cover a wide range of values of interfacial tension (9.1–46.5 mN/m). The systems were kerosene–water, toluene–water, n-butyl acetate–water and butanol–water. First three systems were recommended by the European Federation of Chemical Engineering [25]. The physical properties of these systems are listed in Table 2.

Table 2: Physical properties of chemical systems used

Physical property	Kerosene–water	Toluene–water	n-Butyl acetate–water	n-Butanol–water
ρ_c (kg/m³)	998	998.2	997.6	985.6
ρ_d (kg/m³)	804	864	880	846
μ_c (mPa s)	1.00	0.963	1.0274	1.42

μ_d (mPa s)	1.66	0.586	0.734	3.36
σ (mN/m)	46.5	35.4	13.5	1.9

Technical grade solvents were used at least 99.5 wt% purity as the dispersed phase, and the distillated water was used as the continuous phase. All experiments were carried out at the temperature up to 20 °C. The operating conditions of some of the experiments are shown in Table 3.

Table 3: The operating conditions and hold-up of the dispersed phase for some of experiments

Run	System	Af (cm/s)	Q_c (l/h)	Q_d (l/h)	Holdup
1	n-Butyl acetate–water	0.45	1.1	1.1	0.20
2	n-Butyl acetate–water	0.60	1.1	1.1	0.19
3	n-Butyl acetate–water	0.80	1.1	1.1	0.16
4	n-Butyl acetate–water	0.95	1.1	1.1	0.15
5	n-Butyl acetate–water	1.12	1.1	1.1	0.14
6	Butanol–water	0.45	1.1	1.1	0.17
7	n-Butyl acetate–water	0.80	1.1	2.1	0.22
8	n-Butyl acetate–water	0.80	1.1	3.6	0.24
9	n-Butyl acetate–water	0.80	1.1	5.0	0.28
10	n-Butyl acetate–water	0.80	1.1	7.3	0.30
11	Butanol–water	0.80	1.1	1.1	0.14
12	Butanol–water	0.80	1.1	2.1	0.19
13	n-Butyl acetate–water	0.80	7.3	1.1	0.17
14	Butanol–water	0.80	1.1	1.1	0.11
15	Butanol–water	0.80	2.1	1.1	0.12
16	Butanol–water	0.80	3.6	1.1	0.12
17	Kerosene–water	1.10	1.1	1.1	0.19
18	Kerosene–water	1.30	1.1	1.1	0.17

Procedure

Initially, the whole column was filled with the water which is considered as the continuous phase. The dispersed phase was then entered to column. Sufficient time was provided for reaching the steady state conditions. The drop diameters were measured by photographic method (Nikon D5000 digital camera). Six regions of the active part of column were selected for taking the photographs and calculating the mean drop diameter. To determine the size of the drops, the recorded photos were analyzed by AutoCAD software. In this work for example, three photos of droplets have been shown in Fig. 2. For each experimental condition more than 300 drops were analyzed to guarantee the statistical significance of the determined mean drop size and drop size distribution.

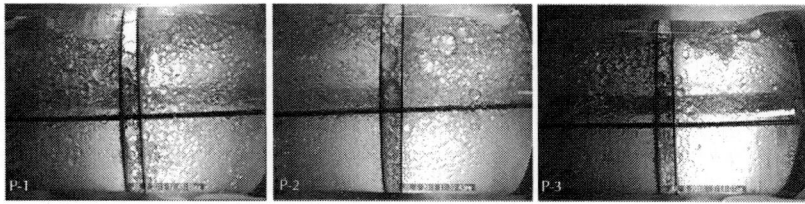

Figure 2: Three photos of droplets for the steady state conditions.

RESULTS AND DISCUSSION

The drop size in the pulsed columns depends upon the drops breakage to smaller drops and drops coalescence to bigger drops [26]. The breaking of drops may be caused by turbulence produced by the pulses, the drops flow through the plate holes, and the hitting of drops with the wall of the columns and the perforated plates [27]. When drops strike each other coalescence of the drops is expected. The breakage and coalescence of the drops produces a range of drop sizes .Thus a mean value of drop sizes and the way these drop sizes are distributed is meaningful in such circumstances.

In this study, three operational parameters were examined including pulse intensity and volumetric flow rates of continuous and dispersed phases. Pulse intensity is defined as the product of pulse amplitude, a(cm) and pulse frequency, f (s^{-1}). The mean diameter is taken as the Sauter mean or volume surface mean diameter, d_{32} defined by the following expression [28]:

$$d_{32} = \left(\frac{\sum\limits_{i=1}^{n} n_i d_i^3}{\sum\limits_{i=1}^{n} n_i d_i^2} \right)$$

(2)

where d_{32} is the Sauter or volume–surface mean diameter, n_i is the number of droplets of mean diameter d_i within a narrow size range i.

The main operating parameters found to affect mean drop size of the column are the pulse intensity and phase flow rates. The effect of pulse intensity on Sauter mean drop diameter is shown in Fig. 3 for some experiments. Drops produced from the high interfacial tension systems (kerosene–water and toluene–water) are larger than those produced from the medium and lower interfacial tension systems (n-butyl acetate–water and butanol–water). It can also be found that the effect of pulse intensity on mean drop size of kerosene–water and toluene–water systems is larger than that of n-butyl acetate–water and butanol–water. That is why the breakup of dispersed phase drops into smaller ones is limited for the latter systems due to their own lower interfacial tension.

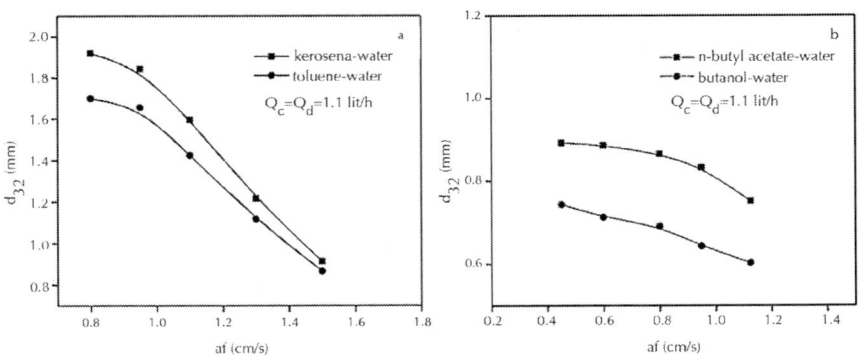

Figure 3: Effect of pulse intensity on mean drop size.

The effect of dispersed phase flow rate on mean drop size is shown in Fig. 4 for some experiments. Increasing the flow rate of dispersed phase leads to increase the drop size. As it is shown in Fig. 4, the effect of dispersed phase flow rate is weaker than the effect of pulse intensity. The increase in droplet size may be attributed to an increase in the coalescence rate due to the larger holdup values that are observed as the dispersed phase flow rate increases.

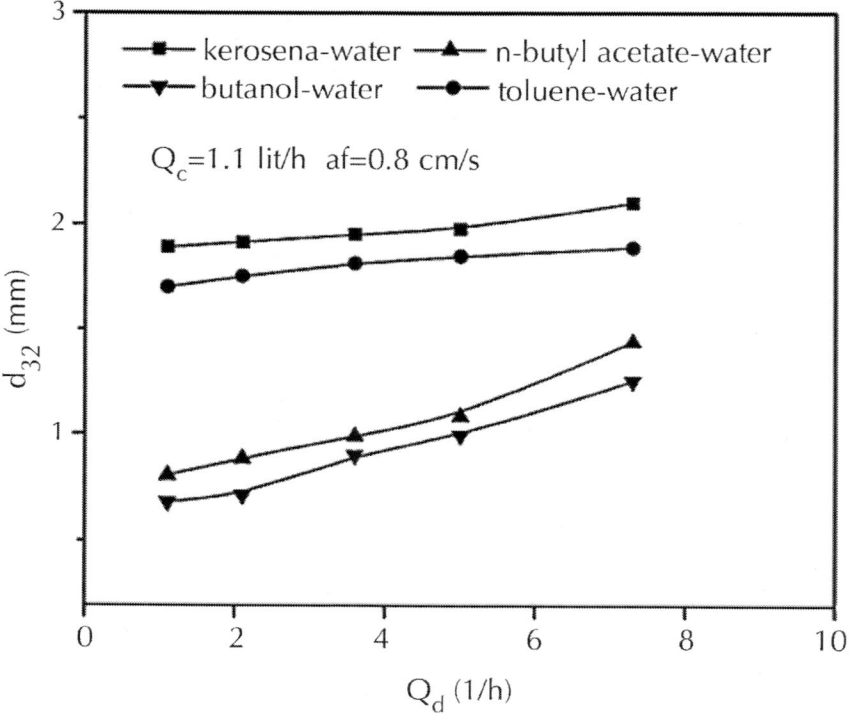

Figure 4: Effect of pulse intensity on mean drop size.

The effect of continuous phase flow rates on Sauter mean drop size is illustrated in Fig. 5 for some experiments. The mean drop diameter slightly increases with an increase in continuous phase flow rates. Increasing the continuous phase flow rates is expected to increase residence time of drops due to the reduction in slip (relative) velocity between the drops and continuous phase.

Increasing the residence time of drops enhances the probability for drop coalescence, leading to larger drops.

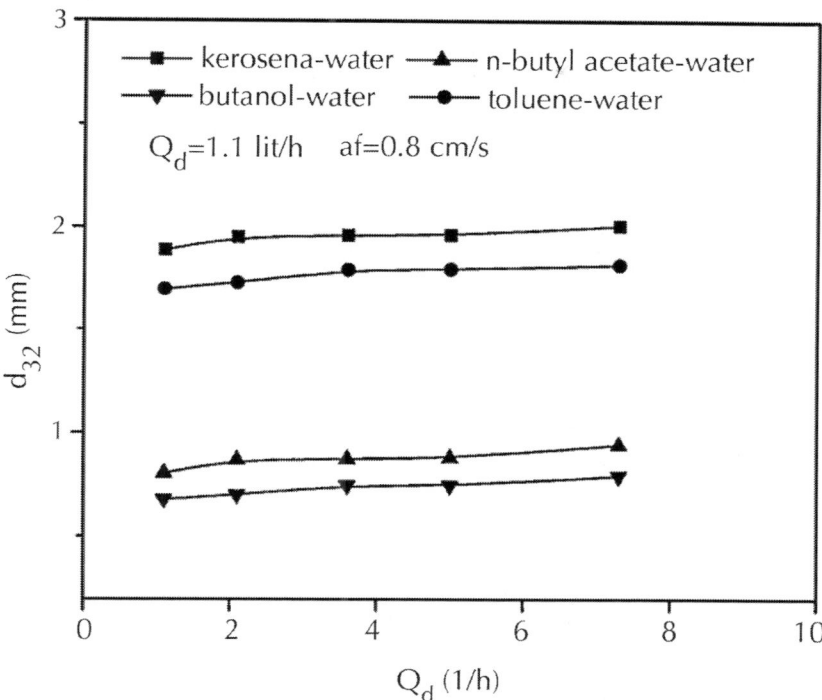

Figure 5: Effect of pulse intensity on mean drop size.

Predictive Correlation for Mean Drop Size

For prediction of mean drop size, a semi empirical correlation (3) is derived in terms of operating parameters and physical properties of the liquid systems by dimensional analysis methods as follows:

$$d_{32} = 2.8$$

$$\times 10^{-4} \left(\frac{1 + Q_c}{Q_d} \right)^{-0.203} \left(\frac{\mu_d}{\mu_c} \right)^{0.025} \left(\frac{\sigma}{(\rho_d \sqrt{af^3 Q_d})} \right)^{0.444}$$

$$(3)$$

The comparison of experimental results with those calculated by Eq. (3) is shown in Fig. 6. This figure indicates that the suggested correlation can estimate the dispersed phase drop size with high accuracy. The AARE for this correlation was found to be about 15.6%.

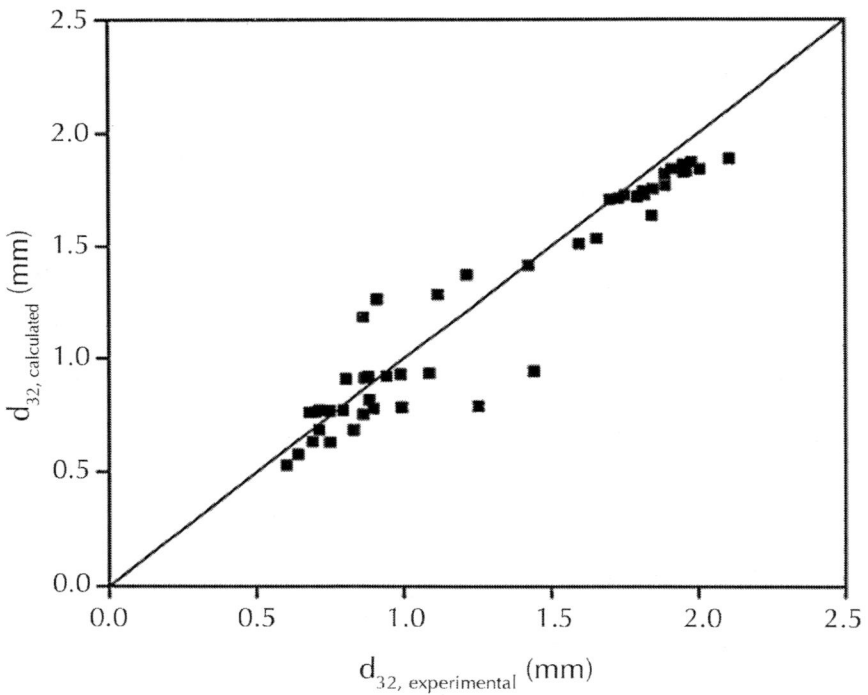

Figure 6: Comparison of experimental mean drop size values with calculated values.

Drop Size Distribution

The effect of the pulse intensity on the drop size distributions for different systems was studied in different operation conditions. For example at constant flow rates of both phases ($Q_c = Q_d = 1.1$ l/h) this effect is shown in Fig. 7. It shows that an increase in the pulse intensity will lead to narrower ranges of distribution for the drop size. The collision energy and turbulence of the system are increased with increasing the pulse intensity, as a result of which, eddies of the liquid stream become smaller. The smaller eddy leads to breakup of the drops whereas the larger ones carry the drops [29]. The importance of the collision energy and the smaller eddies makes the pulse intensity become a key operational parameter from the standpoint of the drop size distribution.

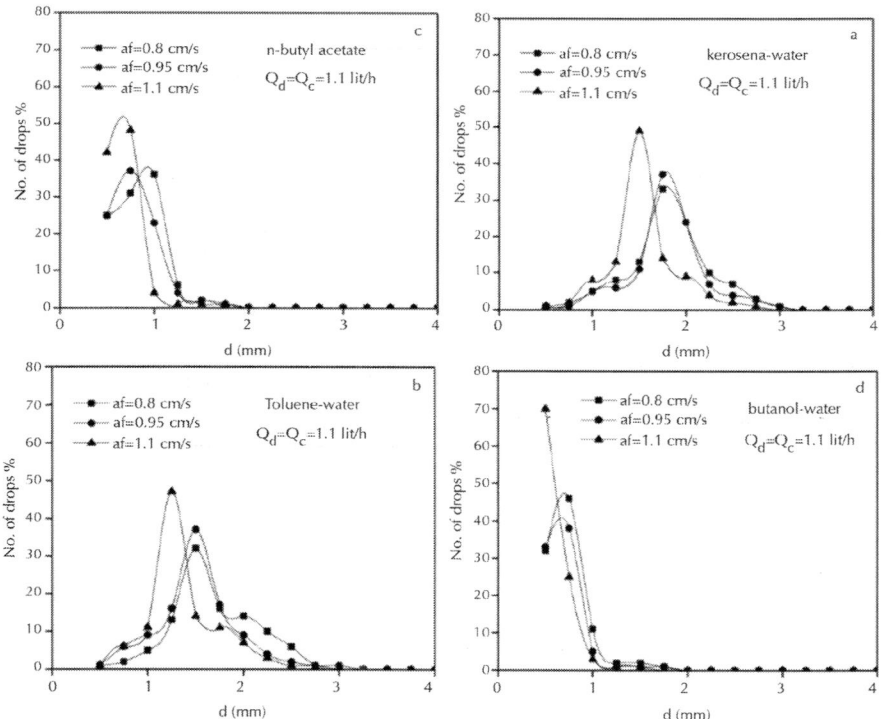

Figure 7: The effect of pulse intensity on the drop size distribution.

At high values of pulse intensity (af), the difference between distribution curves of different liquid systems decreases due to an overriding effect of the pulse intensity against the interfacial tension. At lower the pulse intensity values, the interfacial tension plays a significant role in the shape of distribution curves. The effect of the dispersed phase flow rate on drop size distribution has been shown in Fig. 8 for some experiments. An increase in the dispersed phase flow rates seems to broaden the drop size distribution, because an increase of the dispersed flow rate increases the coalescence frequency.

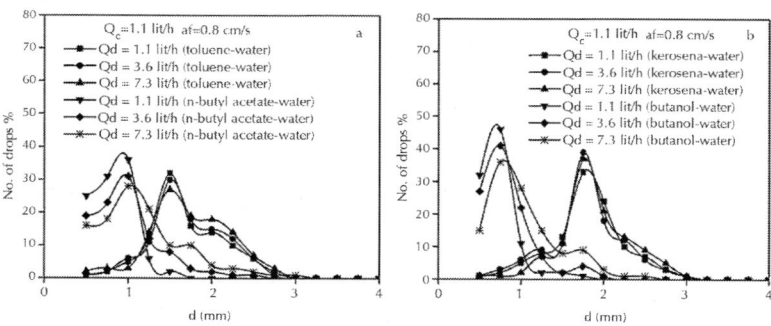

Figure 8: The effect of dispersed phase flow rate on drop size distribution.

The effect of continuous phase flow rate on drop size distribution has been shown in Fig. 9 for some experiments. It shows that, the effect of continuous phase flow rates is weaker than the effect of dispersed phase flow rate. No sensible changes in the trend of the curves are evident, because all flow rates have almost the same distribution points.

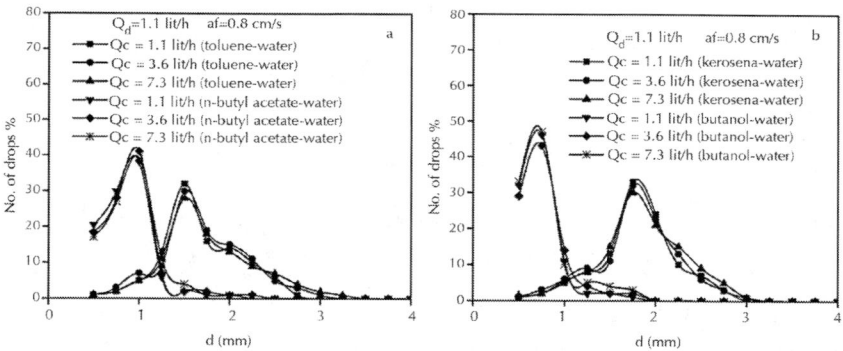

Figure 9: The effect of continuous phase flow rate on drop size distribution.

Predictive Correlation for Drop Size Distribution

Table 3 summarizes the probability distribution functions presented in the literature for liquid–liquid extraction systems. The constants in these functions must be optimized using experimental data. In order to find a suitable correlation, four models of distribution functions were tested for reproduction of the experimental data that has been indicated in Table 4. Among these functions only the normal probability density function (Eq. (4)) was suitable for representing the experimental drop size distributions:

$$P_n(d) = \left(\frac{1}{\sqrt{2\pi d\alpha}}\right) \exp\left[-\left(\frac{(\ln d - \beta)}{\sqrt{2}\alpha}\right)^2\right]$$

(4)

In which d is the drop diameter and α and β are parameters to be fitted. It should be emphasized that the log-normal function was already found to be adequate for describing drop size distributions in extraction columns [19] and [30].

Table 4: Probability distribution functions appear in the literature for liquid–liquid extraction systems

Name	Formula	Reference
Log-normal	$P_n(d) = \dfrac{1}{\sqrt{2\pi d\alpha}} \exp\left[-\left(\dfrac{\ln d - \beta}{\sqrt{2}\alpha}\right)^2\right]$	[18], [19] and [30]
Normal	$P_n(d) = \dfrac{1}{\sqrt{2\pi\alpha}} \exp\left[-\left(\dfrac{d - \beta}{\sqrt{2}\alpha}\right)^2\right]$	[19]
Log-normal	$P_n(d) = \dfrac{\alpha}{d\sqrt{\pi}} \exp\left[-\left(\alpha\ln\dfrac{d}{dq}\right)^2\right]$	[31]
Gamma	$P_n(d) = \dfrac{\beta}{\Gamma(\alpha + 1)} d^\alpha \exp(-\beta d)$	[31]

| Weibull | $P_n(d) = \alpha\beta d_i^{\alpha-1}\exp(-\beta d_i^\alpha)$ | [32] |

The probability density has been considered as a number density defined as the ratio of number of drops of a specific diameter ($d_i \pm d/2$) to the total number of drops. Where d_i is the drop diameter (mm), the two fitted parameters are given in terms of the following dimensionless numbers:

$$\alpha = 0.399\left(1 + \frac{Q_c}{Q_d}\right)^{0.003}\left(\frac{\sigma}{(\rho_d\sqrt{af^3 Q_d})}\right)^{-0.070}$$

$$\times\left(\frac{\mu_c}{(\rho_d\sqrt{af Q_d})}\right)^{0.025}\left(\frac{\mu_d}{(\rho_d\sqrt{af Q_d})}\right)^{0.061} \tag{5}$$

$$\beta = 0.22\left(1 + \frac{Q_c}{Q_d}\right)^{0.014}\left(\frac{\sigma}{(\rho_d\sqrt{af^3 Q_d})}\right)^{0.576}$$

$$\times\left(\frac{\mu_c}{(\rho_d\sqrt{af Q_d})}\right)^{-0.542}\left(\frac{\mu_d}{(\rho_d\sqrt{af Q_d})}\right)^{0.006} \tag{6}$$

where Q_c and Q_d are volumetric flow rates of the continuous and dispersed phases, af is the pulse intensity, μ_d and ρ_d are viscosity and density of the dispersed phase, σ is the interfacial tension, μ_c is viscosity of the continuous phase. In Eqs. (5) and (6), the parameters of α and β have AARE values of about 13.7 and 4.8%, respectively.

CONCLUSIONS

In this paper drop size and drop size distribution in a horizontal pulsed extraction were studied. The observations revealed that the drop size was decreased and the distribution curve of the drop size was sharpened with an increase in the pulse intensity. The flow rates of the dispersed and continuous phases were found to have almost no significant effect; their effects were negligible as compared to the effect of the pulse intensity. Another main parameter was the interfacial tension. It can also be found that the effect of pulse intensity on drop size and drop size distribution of kerosene–water and toluene–water systems was larger than that of n-butyl acetate–water and butanol–water. That is why an increase in the interfacial tension intensifies the tension energy against the collision and shear energies, as a consequence of which, the drop size will increase. Based on the results of experimentation a semi empirical correlation for the estimation of drop size was established which proves to be in good agreement to the experimental findings. This correlation had average absolute relative error (AARE) values of about 15.6%. In order to find a predictive correlation for drop size distribution, four models of distribution functions were tested. The normal probability density function is the only suitable way for representing the experimental drop size distributions with an AARE of 13.7 and 4.8% for their optimized constant.

REFERENCES

1. A. Warade, R. Gaikwad, R. Sapkal, V. Sapkal, Simulation of multistage countercurrent liquid–liquid extraction, Leo. J. Sci. 20 (2011) 79–94.

2. V.S. Kislik, Chapter 13–Advances in development of solvents for liquid–liquid extraction, Solvent Extraction Classical and Novel Approaches, Elsevier, 2012, pp. 451–481.

3. F. Bendebane, L. Bouziane, F. Ismail, Liquid–liquid extraction of naphthalene: application of a mixture design and optimization, Energy Procedia 36 (2013) 1241–1248.

4. J.Y. Lee, J.R. Kumar, J.S. Kim, H.K. Park, H.S. Yoon, Liquid–liquid extraction/ separation of platinum(IV) and rhodium(III) from acidic chloride solutions using tri-iso-octylamine, J. Hazard. Mater. 168 (2009) 424–429.

5. T.A. Todd, J.D. Law, R.S. Herbet, Hazardous and Radioactive Waste Treatment Technologies Handbook, in: C.H. Oh (Ed.), CRC Press, Boca Raton, 2001.

6. M. Attarakih, H.J. Bart, Solution of the population balance equation using the differential maximum entropy method (DMaxEntM): an application to liquid extraction columns, Chem. Eng. Sci. 108 (2014) 123–133.

7. A.A. Hussain, T.-B. Liang, M.J. Slater, Characteristic velocity of drops in a liquid– liquid extraction pulsed sieve plate column, Chem. Eng. Res. Des. 66 (1988) 541–554.

8. R.L. Yadav, A.W. Patwardhan, Design aspects of pulsed sieve plate columns, Chem. Eng. J. 138 (2008) 389–415.

9. M. Lorenz, H. Haverland, A. Vogelpohl, Fluid dynamics of pulsed sieve plate extraction columns, Chem. Eng. Technol. 13 (1990) 411–422.

10. A. Ijaz, A. Shafeeq, A. Muhammad, S.S. Daood, Optimized values for a sieve plate pulsed column for acetic acid, water and kerosene system, J. Qual. Technol. Manage. 6 (2010) 119–133.

11. A.J. Melnyk, Hydrodynamic behavior of a horizontal pulsed solvent extraction column, part 1:flow characterization, throughput capacity and holdup, Can. J. Chem. Eng. 70 (1992) 417–425.

12. A.A. Hussain, T.B. Liang, M.J. Slater, Characteristic velocity of drops in a liquid– liquid extraction pulsed sieve plate column, Chem. Eng. Res. Des. 66 (1988) 541–554.

13. H. Brauer, D. Sucker, Biological waste water treatment in a high efficiency reactor, Ger. Chem. Eng. 2 (1979) 71–86.

14. J. Prochazka, M.M. Hafez, The analysis of the dynamic effects in vibrating and pulse plate extraction columns, Collect. Czech. Chem. Commun. 37 (1972) 3725–3734.

15. D.H. Logsdail, J.D. Thornton, Developments in horizontal pulsed contactors for liquid–liquid extraction processes, J. Nucl. Energy 1 (1959) 15–24.

16. V.M. Vdovenko, S.M. Kulikov, Hydrodynamics and mass transfer of a horizontal pulsed column, Radiokhimiya 8 (1966) 525–533.

17. S. Maa, S. Wollny, A. Voigt, M. Kraume, Experimental comparison of measurement techniques for drop size distributions in liquid/liquid dispersions, Exp. Fluids 50 (2011) 259–269.

18. N.S. Oliveira, D.M. Silva, M.P.C. Gondim, M.B. Mansur, A study of the drop size distributions and hold-up in short Kühni columns, Braz. J. Chem. Eng. 25 (2008) 729–741.

19. E. Moreira, L.M. Pimenta, L.L. Carneiro, R.C.L. Faria, M.B. Mansur, J.C.P. Ribiero, Hydrodynamic behavior of a rotating disc contactor under low agitation conditions, Chem. Eng. Commun. 192 (2005) 1017–1035.

20. S. Soltanali, Y. Ziaie-Shirkolaee, Experimental correlation of mean drop size in rotating disc contactors (RDC), J. Chem. Eng. Jpn. 41 (2008) 862–869.

21. A.M.I. Al-Rahawi, New predictive correlations for the drop size in a rotating disc contactor liquid–liquid extraction column, Chem. Eng. Technol. 30 (2007) 184–192.

22. R.L. Yadav, A.W. Patwardhan, Design aspects of pulsed sieve plate columns, Chem. Eng. J. 138 (2008) 389–415.

23. M. Gholam Samani, A. Haghighi Asl, J. Safdari, M. Torab-Mostaedi, Drop size distribution and mean drop size in a pulsed packed extraction column, Chem. Eng. Res. Des. 90 (2012) 2148–2154.

24. M.R. Usman, H. Sattar, S.N. Hussain, H. Muhammad, A. Asghar, W. Afzal, Drop size in a liquid pulsed sieve-plate extraction column, Braz. J. Chem. Eng. 26 (2009) 677–683.

25. T. Misek, R. Berger, J. Schröter, Standard test systems for liquid extraction studies, EFCE Publ. Ser. 46 (1985) .

26. S. Schlauch, Modeling and simulation of drop size distributions in stirred liquid–liquid systems, Ph.D. Thesis, Technischen Universitat Berlin, Berlin, 2007.

27. K. Sreenivasulu, D. Venkatanarasaiah, Y.B.G. Varma, Drop size distributions in liquid pulsed columns, Bioprocess Eng. 17 (1997) 189–195.

28.   L. Boyadzhiev, M. Spassov, On the size of drops in pulsed and vibrating plate extraction columns, Chem. Eng. Sci. 37 (1982) 337–340.

29.   H.T. Chen, S. Middleman, Drop size distribution in agitated liquid–liquid systems, AlChE J. 13 (1967) 989–995.

30.   C. Tsouris, R. Ferreira, L.L. Tavlarides, Characterization of hydrodynamic parameters in a multistage column contactor, Can. J. Chem. Eng. 68 (1990) 913–923.

31.   L.M. Rincon-Rubio, A. Kumar, S. Hartland, Drop size distribution and average drop size in a Wirz extraction column, Chem. Eng. Res. Des. 72 (1994) 493–502.

32.   L.S. Tung, R.H. Luecke, Mass transfer and drop sizes in pulsed-plate extraction columns, Ind. Eng. Chem. Process Des. Dev. 25 (1986) 664–673.

Pressure Drop Estimation in Horizontal Annuli for Liquid–gas 2 Phase Flow: Comparison of Mechanistic Models and Computational Intelligence Techniques

Reza Ettehadi Osgouei[a], A. Murat Ozbayoglu[b],
Evren M. Ozbayoglu[a], Ertan Yuksel[b], and
Aydın Eresen[c]

[a]The University of Tulsa, Department of Petroleum Engineering, Tulsa, OK 74104, USA

[b]TOBB University of Economics and Technology, Department of Computer Engineering, Ankara 06560, Turkey

[c]Texas A&M University, Department of Electrical and Computer Engineering, College Station, TX 77843, USA

ABSTRACT

Frictional pressure loss calculations and estimating the performance of cuttings transport during underbalanced drilling operations are more difficult due to the characteristics of multi-phase fluid flow inside the wellbore. In directional or horizontal wellbores, such calculations are becoming more complicated due to the inclined wellbore sections, since gravitational force components are required to be considered properly. Even though there are numerous studies performed on pressure drop estimation for multiphase flow in inclined pipes, not as many studies have been conducted for multiphase flow in annular geometries with eccentricity. In this study, the frictional pressure losses are examined thoroughly for liquid–gas multiphase flow in horizontal eccentric annulus.

Pressure drop measurements for different liquid and gas flow rates are recorded. Using the experimental data, a mechanistic model based on the modification of Lockhart and Martinelli [18] is developed. Additionally, 4 different computational intelligence techniques (nearest neighbor, regression trees, multilayer perceptron and Support Vector Machines – SVM) are modeled and developed for pressure drop estimation.

The results indicate that both mechanistic model and computational intelligence techniques estimated the frictional pressure losses successfully for the given flow conditions, when compared with the experimental results. It is also noted that the computational intelligence techniques performed slightly better than the mechanistic model.

INTRODUCTION

Two-phase flows in different geometries are of importance in boilers, nuclear reactors, oil production, drilling operation, electronic cooling, and various types of chemical reactors. Because of its extensive application in different fields, several studies were performed for flow patterns identification, void fraction prediction and pressure drop estimation in annular geometries using different data prediction models, such as linear and nonlinear regression and Artificial Neural Networks [22], as well as mechanistic models [38].

In drilling industry, frictional pressure prediction in wellbore is one of the most critical factors, at which drillstring configuration should be taken into account [10] and [11] especially in any underbalance operation and detection of kick by intelligent drillpipe [15] and [16], because it is used as input for determining numerous other key hydraulics parameters, including the equivalent circulated density (ECD). The aerated fluids have a potential to increase rate of penetration, minimize formation damage, minimize lost circulation, reduce drill pipe sticking and therefore, assist in improving the productivity. Recently, the technology of drilling using aerated fluids has reached even in the area of offshore drilling. The use of compressible drilling fluids in offshore technology has found applications in old depleted reservoirs and in the new fields with special drilling problems. Both hydraulic behavior and mechanism of cutting transport of the drilling fluids formed by gas–liquid mixture are not fully understood yet, especially there is a large uncertainty in prediction of frictional pressure losses. So, the study of annular two phase flow is still continuing because of the need for increasing the accuracy of predicted models for new application areas.

Two-phase flow in horizontal pipes has been on the focus for a lot of theoretical and experimental studies for some time, as a result several different models are developed. These models are generally categorized into two main groups, empirical models and mechanistic models. Initially, the empirical models treated the two-phase flow as one-phase using simplified versions of the actual flow configuration, furthermore, the flow pattern types are not considered in these models. The most notable studies of this type are Wallis [37], Lockhart and Martinelli [18], Duns and Ros [9]. After these initial models, flow pattern identification became the area of concentration; different empirical and mechanistic models were developed during that time. The studies of Dukler et al. [8] and

Beggs and Brill [3] used equations based on experimental data for identifying flow patterns. Mechanistic models used corresponding equations constructed for each different flow pattern, once the flow pattern was identified. Taitel and Dukler [31], Barnea [2], Xiao et al. [38], Petalas and Aziz [25] were among the developed important mechanistic models. Transition boundaries between the flow patterns were investigated based on conservation equations by Taitel and Dukler [31]. In this study, equilibrium condition for stratified flow was assumed. After that, the Lockhart and Martinelli parameter was used in order to determine equilibrium liquid holdup. The Kelvin–Helmholtz inviscid theory was modified in order to predict the initiation of slugs. The transition of intermittent to annular flow is assumed to be dependent only on liquid level. Jeffrey's theory for wave initiation is used to determine the transition of stratified smooth to stratified wavy flow pattern. They investigated turbulent and buoyant forces acting on a gas pocket for the boundary between dispersed bubble flow and intermittent flow. Dimensionless parameters were also developed to express the transition conditions. Barnea [2] studied the transition mechanisms for each individual boundary and proposed a unified model. The developed mechanisms were applicable for the whole range of pipe inclinations. The dimensionless maps were developed to incorporate the effects of flow rates, fluid properties, and pipe size and inclination angle. This model was verified with the conducted experiments. Xiao et al. [38] developed a comprehensive mechanistic model for two-phase flow in horizontal and near horizontal pipes. Taitel and Dueler's [31], and Barnea [2]dimensionless groups were used to predict flow pattern transitions.

Although there are a lot of studies regarding with two-phase flow in circular pipes, limited investigation have been conducted for two-phase flow through annulus. Some examples are; Sadatomi et al. [27], Caetano et al. [6], Hasan and Kabir [13]. Sadatomi et al. [27] were most probably the first one to develop the flow pattern maps for the flow through annuli. Hasan and Kabir [13] recognized four major flow regimes-bubbly, slug, churn and annular from the estimated void fraction for air–water systems. In case of bubbly flow, they found out that the terminal rise velocity was not affected significantly by either the variation in the inner tube diameter or the channel deviation from the vertical. Similarly, in this regime they concluded that the void fraction was not affected by inclination angle. Caetano et al. [6] carried out experimental and theoretical study of upward gas–liquid flow through vertical concentric and eccentric annuli with air–water and air–kerosene mixtures. They identified flow patterns and developed flow pattern maps based on visual observations in conducted experiments. Moreover, they developed mechanistic models for prediction of average liquid holdup and pressure drop for each flow pattern in concentric and eccentric annular geometries. Sunthankar [30] modified Taitel and Dukler [31] transition equations for determining the flow patterns for annular geometries by using the definition of hydraulic diameter. He also compared the estimated results by experimental results. Lage et al. [17] experimentally and theoretically studied two-phase fluid flow in horizontal and inclined annulus. Equations from Taitel and Dukler [31] were used to determine flow patterns. Ozbayoglu and Omurlu [24] formed a mechanistic model to determine the flow patterns and to calculate the frictional pressure losses of gas–liquid mixture fluid in horizontally located annular. Based on experimental observations, Osgouei et al. [21], Osgouei et al. [12] developed a mechanistic model for determining the total pressure losses and volumetric distribution of two phase fluids flow within the inclined wellbore for a particular drilling condition. Their proposed model is reasonably accurate for estimating the frictional pressure losses when compared with the measured values.

Since mechanistic models were mostly concerned about identifying the correct flow patterns, other techniques such as Artificial Neural Networks (ANN) were approached for estimating the flow rates. However, the amount of studies and models developed for two-phase flow in horizontal annulus has been limited, some notable ones include Ternyik et al. [32] and [22]. These studies used experimental data sets provided by other researchers. Ozbayoglu and Ozbayoglu [23] provided several different neural network models for flow pattern prediction and pressure drop estimation using flow rates and fluid parameters and the implemented models provided promising results. Alizadehdaknel et al. [1] compared the performance of CFD models with ANN in pressure drop estimation in multiphase flow and observed that CFD results were more accurate than ANN [1].

There are also other computational intelligence models developed in similar problems. Timung [34] used a probabilistic neural network for identifying flow patterns through different circular micro-channel settings (Timung [34]), using water–gas flow. Zhao developed an SVM model in predicting pressure drop for cyclone separators [39].

Research is still ongoing in this field, and new results are obtained as more researchers and drilling companies are interested in achieving higher precision in estimating the flow behavior.

The mechanistic models, which consider the inter-phase momentum transfer, are the two-fluid or multi-fluid models. In this approach, separate momentum equations for gas, liquid, and droplets are written. In addition, a closure relationship for inter-phase drag forces is assumed, which incorporates the slippage between the phases. The multi-fluid approach has been applied in the commercial pipeline simulator such as OLGA [4].

Belt et al. [5] tested the performance of two commercial multiphase pipe flow simulators, OLGA 5.3 and LedaFlow, on laboratory and field data. Although the results were comparable by the laboratory data, two weak points have been identified: an over prediction of the liquid hold-up in stratified flow and a relative error up to 70% on the pressure-gradient in intermittent flow.

The main objective of current study is propose the fast and simplified calculation methodology with reasonable accuracy, so that complex simulator results may not be required.

In order to identify the drilling parameters those have the major influence on the horizontal drilling process and to define the flow pattern types and boundaries, the experimental study which consists of two-phase air–water experiments in large-scale annuli (2.91" × 1.86", approximately 21' long) in horizontal direction have been conducted. Experiments were carried out without drill pipe rotation. For each experiment, average void fractions and total pressure drop along the flow loop was measured.

Accompanied by the experimental study, theoretical analysis to predict frictional pressure losses for horizontal air–water flow in annuli was carried out. A model developed by Lockhart and Martinelli [18] for pressure loss prediction in horizontal pipes was modified for flow in annuli by taking to account the annular geometry with inner pipe eccentricity. Meanwhile, different machine learning models are developed for pressure drop estimation. In this study, nearest neighbor, regression trees, multilayer perceptron and Support Vector Machines are chosen as computational intelligence models. The dimensionless parameters, superficial Reynolds numbers, were developed in the computational intelligence models to consider the effect of all factors on pressure drop.

The acquired experimental data were used to evaluate both the modified pressure drop prediction model and artificial intelligence models.

EXPERIMENTAL SETUP

During the experiments, 398 different flow runs with different liquid and gas flow rates are implemented and the corresponding data are collected. The working environment and range of input values were tabulated in Table 1.

Table 1: Test matrix for gas–liquid two phase flow tests

	Minimum	Maximum
Range water annular velocity (ft/s)	1	10
Range gas annular velocity (ft/s)	1	120
Range annular pressure (Psi g)	1	13
Pressure drop (psi/ft)	0.0063	0.7114
Temperature (°C)	25	35
Eccentricity ratio	0.623	0.623

Comprehensive experiments have been conducted in Middle East Technical University Petroleum and Natural Gas Engineering Department Multiphase Flow Loop. The flow loop consists of annular test section with length of 21 ft. and 2.91 in. inner diameter transparent acrylic casing with a 1.85 in. outer diameter inner drillpipe (Figs. 1a and 1b). The liquid collected in the tank was pumped and circulated through the loop. Two centrifugal pumps (maximum capacity of 250 gpm) were used periodically with a Fisher control valve to have a controlled circulation of liquid through the loop. Also, gas was injected into the flow loop using a compressor of 100 scfm, and the gas flow rate was measured and controlled by a mass flow meter and a pneumatic flow controller, respectively.

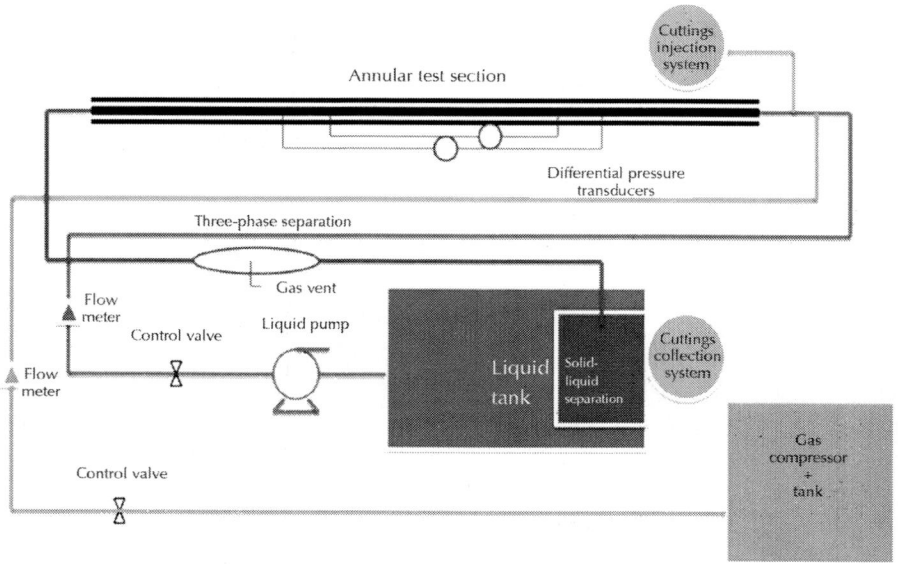

Figure 1a: Schematic of experimental setup.

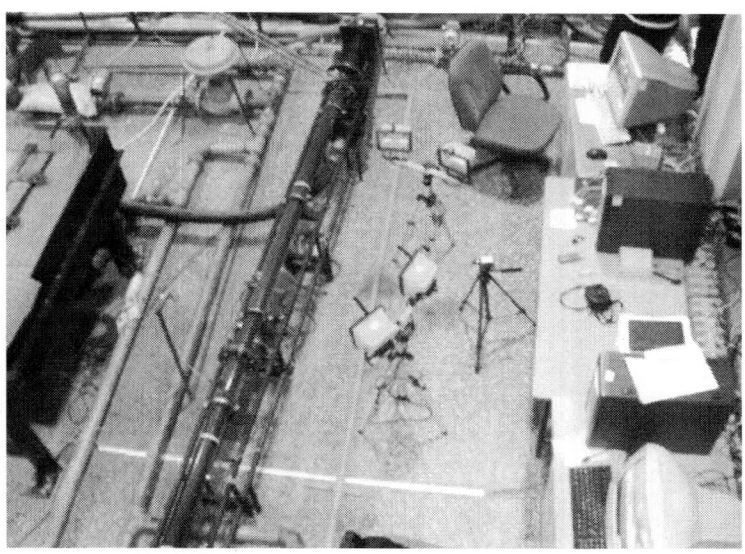

Figure 1b: METU-PETE multiphase flow loop.

The experiments were performed in an eccentric annulus using water–air mentioned positions, without inner pipe rotation. The standard experimental procedure adapted was as follows: Using centrifugal pump, the liquid was pumped at a constant flow rate in the rage of 0–250 gpm. Then, the air was injected into the annular test section through a compressor with the working capacity of 0–50 psi and a flow rate rage 0–100 scfm. Once both the air and liquid flow rates were stabilized, the data acquisition was activated to record flow rates, pressures at critical points, pressure drop inside the test section.

MECHANISTIC MODEL

The mechanistic model developed for two phase liquid–gas flow is a modified version of Lockhart and Martinelli [18] method. Lockhart and Martinelli [18] suggested a model for estimating the pressure gradient in the two-phase pipe flow. In this method, the total pressure gradient was obtained by the multiplying liquid or the gas surface pressure gradient by dimensionless pressure gradient parameters (φ_L and φ_G). Mathematical description of this method is given below.

$$\frac{\Delta P}{\Delta L} = \phi_L^2 \left(\frac{\Delta P}{\Delta L}\right)_{SL} = \phi_G^2 \left(\frac{\Delta P}{\Delta L}\right)_{SG} \tag{1}$$

where (φL), (φG), $\left(\dfrac{\Delta P}{\Delta L}\right)_{SG}$ and $\left(\dfrac{\Delta P}{\Delta L}\right)_{SL}$ are liquid, gas dimensionless pressure gradient parameters and gas, liquid surface pressure gradients, respectively. Generally, gas and liquid surface pressure gradients are associated with Lockhart–Martinelli parameter (X) as in Eq. (2).

$$X = \sqrt{\frac{\left(\frac{\Delta P}{\Delta L}\right)_{SL}}{\left(\frac{\Delta P}{\Delta L}\right)_{SG}}} \tag{2}$$

Where

$$\left(\frac{\Delta P}{\Delta L}\right)_{SL} = \frac{2}{(D_o - D_i)g} C_L \left[\frac{1488\, v_{SL}\rho_L(D_o - D_i)}{\mu_L}\right]^{-n} \rho_L v_{SL}^2 \tag{3}$$

Lockhart and Martinelli [18]

$$\left(\frac{\Delta P}{\Delta L}\right)_{SG} = \frac{2}{(D_o - D_i)g} C_G \left[\frac{1488\, v_{SG}\rho_G(D_o - D_i)}{\mu_G}\right]^{-m} \rho_G v_{SG}^2 \tag{4}$$

Lockhart and Martinelli [18]

$C_L = C_G = 16$ and $m = n = 1$ for laminar flow, $C_L = C_G = 0.046$ and $m = n = 0.2$ for turbulent flow.

Lockhart–Martinelli parameter (X) and liquid dimensionless pressure gradient parameter (φ_L) were calculated for current experimental data by using Eqs. (1), (2), (3) and (4). The relationship equation between Lockhart–Martinelli parameter (X) and liquid dimensionless pressure gradient parameter (φ_L) were modified for current experimental data by using Matlab Curve Fitting Toolbox (Fig. 2) as follows.

$$\varphi_L = 2.717 X^{-0.5901} + 1.085 \tag{5}$$

Eq. (5) was developed based on experimental results. The range of applicability of Eq. (5) is limited to the range of the experimental parameters illustrated in Table 1.

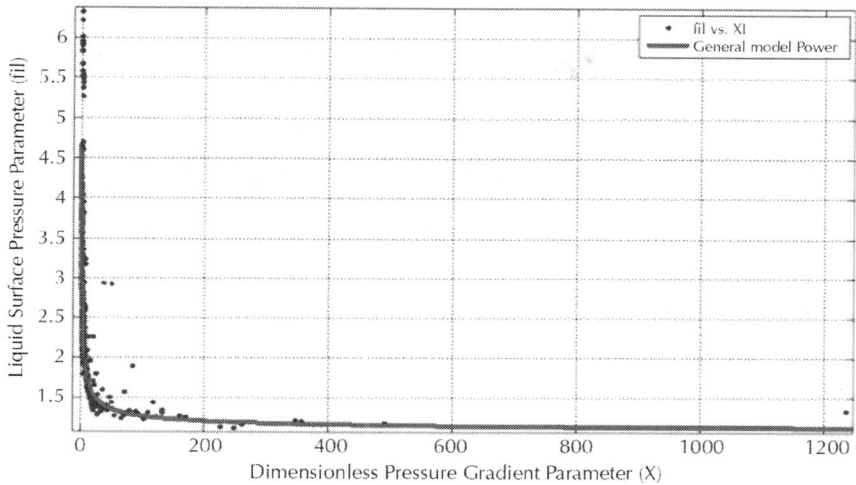

Figure 2: Relationship between liquid dimensionless pressure gradient parameter and liquid surface pressure gradients.

COMPUTATIONAL INTELLIGENCE TECHNIQUES

In this study, 4 different computational intelligence techniques are modeled for pressure drop estimation. Pressure drop values, along with the liquid, gas flow rates used during the experimental data collection process are provided in Table 1. Unlike the mechanistic model, in all of the computational intelligence models used in this paper, Lockard–Martinelli equations are not used, the pressure drop data measured under different settings of liquid and gas flow rates are directly used.

The 4 different computational intelligence models used in this study are nearest neighbor, regression trees, multilayer perceptron and Support Vector Machines – SVM.

Nearest Neighbor Model

Nearest neighbor is one of the simplest classifiers that are used in the machine learning applications. There is no learning involved in this type of classifier, so modeling is extremely fast. However the testing stage is computationally more expensive. The data point that is to be classified or predicted will be compared to every single data point that is introduced before. The output of the data point in the training set that is closest to the test point (nearest neighbor) is chosen as the desired value for the test point. Euclidean distance is used in the determination of the distance between the data points [33]. Nearest neighbor models work better in small dimensions.

Regression Tree

Regression tree is an extension of decision tree where classification of data is performed by selecting the most appropriate separating value of each input feature based on the entropy of the resulting class separation. In decision tree, the choice is determined by finding the input feature and cutoff value for that feature that results in the minimum entropy for the particular classification. The feature value performing the best separation between classes is chosen at each tree level starting at the root. Same procedure is repeated by creating branches (just like in an actual tree) for the separated subsets of the data until all the class data is properly separated, and/or the entropy values for the branches are below a threshold. The details of the underlying decision tree algorithm can be found in Russell and Norvig [26].

The structure of regression tree is very similar to the decision tree, however instead of using entropy as the splitting criteria, variance between similar data points under the same branch is used. The splitting comes to an end when the variance levels under each branch are below an acceptable threshold value. More information about regression trees can also be found in Russell and Norvig [26].

Multilayer Perceptron Neural Network

Multilayer perceptrons are among the most commonly used neural networks for general purpose data forecasting and/or estimation problems [14]. Multilayer perceptrons use a learning algorithm known as Backpropagation where the input signal is fed forward into the network, but the error at the output layer is propagated back to the input layer while the necessary modifications to the network weights are applied so that the error at the output is minimized. As a result of this process, these types of structures are also called as Backpropagation neural networks. These networks have good generalization capabilities and well suited to problems where the relations between input and output are not known and/or clear. Under these circumstances, it is not easy to find a good representing function for the input–output relation, so a neural network mapping between input and output vectors can provide a better outcome for the problem in hand. In neural networks, the sum of error squares at the output is minimized by iteratively modifying the network weights. Interested readers can refer to [14] for a complete analysis of the learning algorithm and detailed model development for such neural networks.

Support Vector Regression

Support Vector Machines (SVM) is an optimized binary classifier that tries to maximize the margin between two classes. When applied to regression problems, a similar approach using Support Vector Regression (SVR) is chosen. SVM framework has been established by Vapnik and Chervonenkis in the early sixties[35], however, it was not until the 1990s, industrial and real world application developers started considering using SVM as a machine learning tool for good generalization and classification [28]. Later, SVR was setup to be used in regression applications, as well [20], [7], [29] and [19].

The SVR formulation is based on the idea that the outputs of the optimized regression function approximation is within ε-neighborhood of the actual data points, however some deviation is allowed for preventing to come up with an infeasible solution by introducing slack-variables (ζ) [36]. The regression function used takes the form as in (6).

$$f(x) = \langle w, x \rangle + b \quad \text{with } w \in X, b \in R$$

(6)

$\langle w, x \rangle$ is defined as the dot product of the weight coefficients and the input vector x. In this particular formulation, linear combination of weights are chosen for the regression function approximation, however, a nonlinear function can also be implemented with the introduction of appropriate kernels used on the input vector x. The SVR function is formulated based on satisfying the following constraints as stated in [36] and [28]:

$$minimize \; \frac{1}{2}\|w\|^2 + C\sum_{i=1}^{l}(\zeta_i + \zeta_i^*)$$

(7)

$$Subject \; to \begin{cases} y_i - \langle w, x_i \rangle - b & \leq \varepsilon + \zeta_i \\ \langle w, x_i \rangle + b - y_i & \leq \varepsilon + \zeta_i^* \\ \zeta_i, \zeta_i^* & \geq 0 \end{cases}$$

where C is a trade-off value determined for how much deviation is tolerated beyond the ε-neighborhood of the solution for some of the data points that end up being more than ε different than the actual output. This is controlled by introducing the slack variables ζ_i (when data point i has an output that is less than $y - \varepsilon$) and ζ_i^* (when data point i has an output that is more than $y + \varepsilon$). When the output is within the ε-neighborhood of the solution, then ζ_i is selected as 0. As a result, the errors of the data points associated within ε-neighborhood are not considered during the regression cost determination. This cost function is shown in Fig. 3. When Fig. 3 is analyzed, one obvious observation is the insignificance of small errors within the ε-neighborhood; these errors do not contribute to the total error associated with the model. The second observation is the linear (not quadratic as in square of error models) increase in error outside the ε-neighborhood. In Fig. 4, the resulting SVM classification model is illustrated where the classifier decision boundary lies in between the support vectors. Here the decision boundary between the two classes are formed by choosing the points closest to the separating line (or hypersurface if number of inputs is more than 2) as the support vectors. With this approach, the interclass distance between the two classes (each data point with the first class is represented with the circles on the right side of Fig. 4, each data point with the second class is represented with the x's on the left side of Fig. 4) is maximized, hence the classification performance of the test and/or production data (out-of-sample data) will have the best possible outcome.

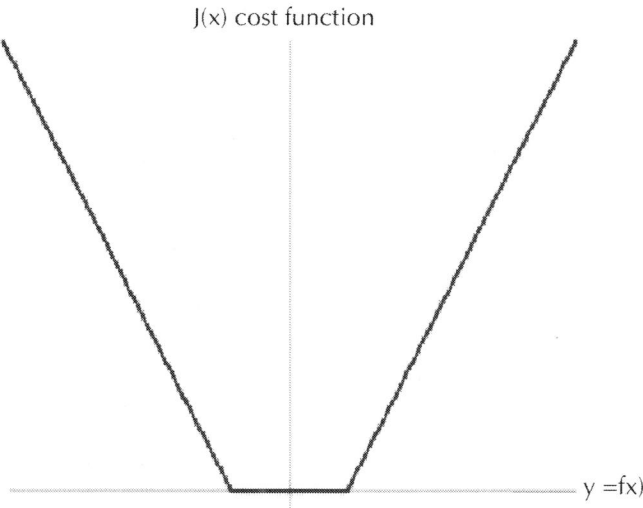

Figure 3: Cost function associated with ε-SVR.

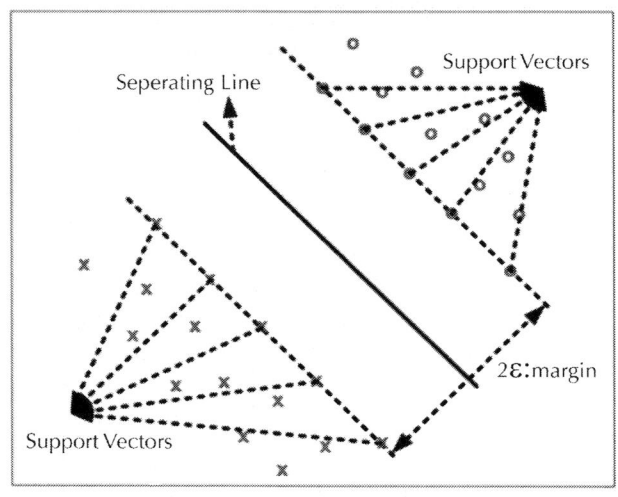

Figure 4: Illustration of how SVM decision boundary (separating line) is chosen.

Model Inputs and Outputs

Same input and output features are created for all computational intelligence models. The superficial Reynolds numbers for both phases are used to simplify and generalize the inputs: measured liquid and gas superficial velocities. This is implemented so that same developed model could be used for different pipe geometries. They are easy to calculate and widely used in solving the multiphase flow problems. Superficial Reynolds number for liquid phase (subscript L) is defined as:

$$N_{Re_L} = \frac{1488 \rho_L V_{SL}(D_o - D_i)}{\mu_L} \tag{8}$$

and superficial Reynolds number for gas phase (subscript G) is expressed as:

$$N_{Re_G} = \frac{1488 \rho_G V_{SG}(D_o - D_i)}{\mu_G} \tag{9}$$

where ρ is density, V is velocity, D_o is wellbore diameter, D_i is pipe diameter, μ is viscosity. Output value is the pressure drop measured between the two ends of the experimental data collection section. The range of the input and output data is tabulated in Table 1 in the Experimental Setup section.

RESULTS AND DISCUSSIONS

The relation between the flow parameters and the friction pressure losses were investigated based on experimental results. As shown in Fig. 5, the friction pressure drop increases with the increase in gas superficial velocity at high liquid velocities, but in low liquid velocities, the change in friction pressure loss is minor by increasing the gas superficial velocity. Also, the friction pressure drop was observed to be increasing with increase in the liquid superficial velocity.

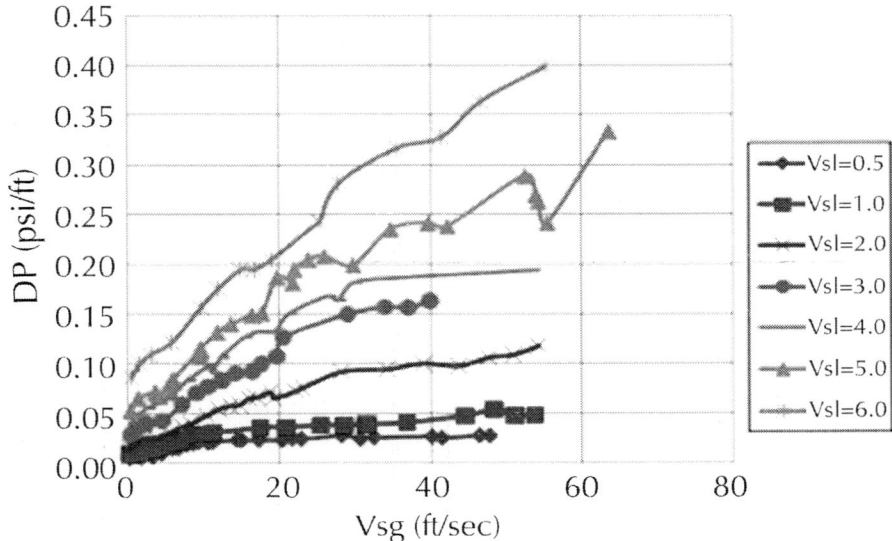

Figure 5: Measured pressure drop vs. gas superficial velocity in horizontal test section ($\theta = 0$) for different water flow rate.

The performance results from the mechanistic model and the computational intelligence models are presented in this section. Two-phase flow pressure drop was estimated though the horizontal annuli by substituting calculated dimensionless pressure gradient parameter (φ_L) by Eq. (5) in Eq. (1). The accuracy of modified Lockhart-Martinelli method was tested by comparing the estimated friction pressure losses using the model and the experimental data obtained from this study and Sunthankar's study [30] (Fig. 6 and Fig. 7). As shown in Fig. 6 and Fig. 7, modified Lockhart-Martinelli model can estimate friction pressure loss with acceptable accuracy for both data sets (with the %20 error). It is also concluded fromFig. 6 and Fig. 7 that the model accuracy is improved by increasing gas and liquid flow rates.

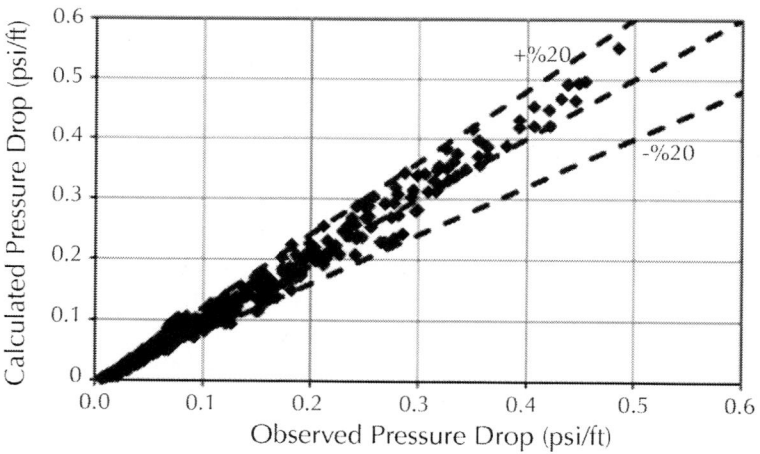

Figure 6: Comparisons between experimental data obtained from current study and estimated friction pressure drop for flow in horizontal annuli by using modified Lockhart-Martinelli Method.

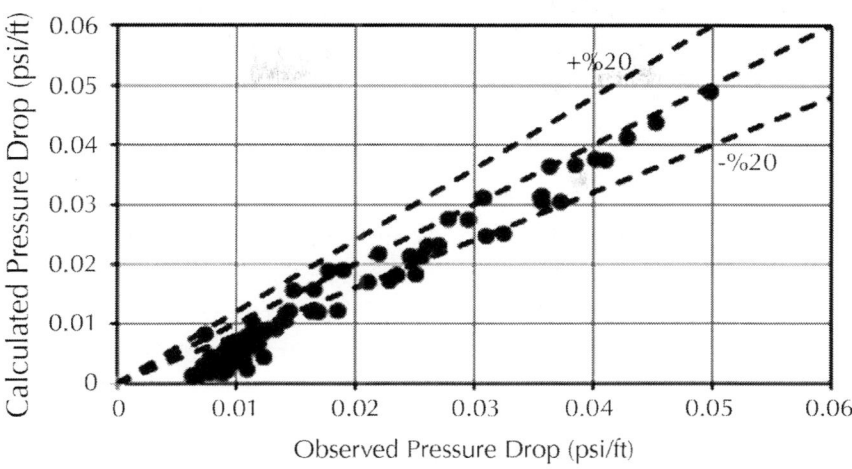

Figure 7: Comparisons between Experimental Data Obtained from Sunthankar [30] and Estimated Friction Pressure Drop for Flow in Horizontal Annuli by Using Modified Lockhart–Martinelli method.

The computational intelligence models were implemented using Matlab which had all the models as part of their library. During the implementation same inputs were used in all 4 models. Also, same training and testing sets were used in all the models. How these models work was explained in the previous section. All these models work as black box systems, as such the user provides the input to the chosen model and the model dynamics (as explained in the previous section) take over and the test data output (pressure drop) is calculated according to the input–output mapping of the training data. Since all 4 models have different internal dynamics for this mentioned mapping, they have different performances.

Table 2 tabulates the performance results for 4 different computational intelligence models developed for liquid–gas 2-phase. The Root Mean Square Error (RMSE), Average Percent Error (APE), Average Absolute Percent Error (AAPE) and correlation coefficient (R) between the estimated values of pressure drop (psi/ft) and the experimental measurements are calculated for the proposed models.

Table 2: Prediction performance of 4 computational intelligence models for 2-phase (liquid–gas) flow (pressure drop (psi/ft) difference between estimated and experimental data)

	RMSE	APE (%)	AAPE (%)	R
Nearest neighbor	0.0301	0.58	9.51	0.9785
Multilayer perceptron	0.0274	0.77	7.42	0.9822
Regression Tree	0.0336	3.18	14.80	0.9731
SVR	0.0173	3.090	7.464	0.9918
Average	0.0271	1.905	9.7985	0.9814

When Table 2 is analyzed, it is clear that SVR and the Multilayer perceptron models provided better performance compared to nearest neighbor and regression tree models. In this particular case regression tree performed the worst, the main reason being the association of the output data and the input data being highly correlated in multi-feature input combination but not so in single feature to output correlation. In such cases, even though the linear or nonlinear regression models can provide good performances, since they can use feature combinations to provide the association of the inputs and outputs, regression trees, by its structural design, can only use one input feature at a time for splitting decision. As a result, even though its performance was acceptable, the outcome was better for other models. SVR model can be chosen as the best model, since from an overall analysis perspective, it had the best combined performance. The error deviation results of SVR are shown in Fig. 8, where the SVR predicted pressure drop (psi/ft) and the experimental results show good alignment. SVR provided the best results of all the models used and developed in this study. However, all computational intelligence models showed all-around satisfactory results in the pressure drop estimation process.

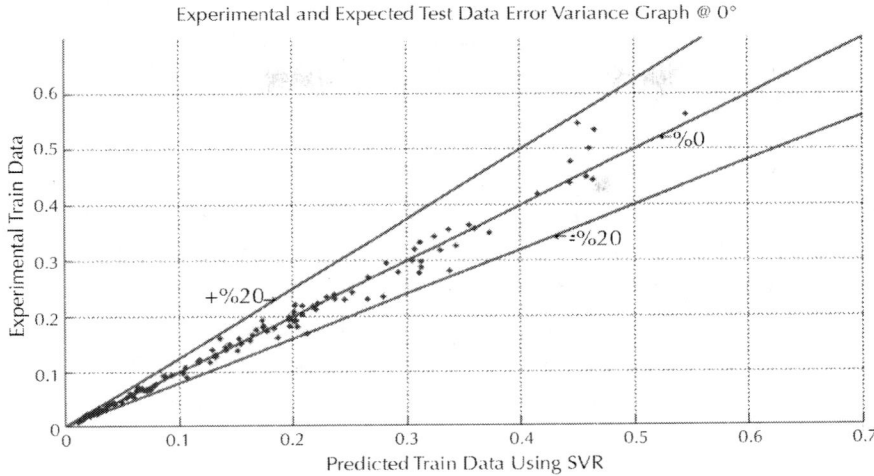

Figure 8: SVR prediction accuracy.

Table 3 demonstrates how the resulting estimated values are distributed when compared to experimental data results. These results also confirm that Support Vector model provided the best overall results due to its error tolerance that was discussed in the previous section.

Table 3: Comparison of prediction results with experimental data

Deviation from experimental data	Number of data points in the selected range for the chosen model				
	Nearest neighbor	Regression tree	MLP	SVR	Average
Less than %5	148	98	196	230	168
In between %5 and %15	176	137	155	135	150.75
In between %15 and %25	49	98	32	23	50.5
In between %25 and %50	23	58	14	5	25
In between %50 and %75	2	5	0	4	2.75
More than %75	0	2	1	1	1

CONCLUSIONS

In this study different mechanistic and computational intelligence were developed for estimating the pressure drop in eccentric horizontal liquid–gas flow. The mechanistic model was designed as a modified version of a previous model in literature. At the same time 4 different computational intelligence models were developed for performance comparison. But the computational intelligence models did not use the Lockhart-Martinelli equations like the mechanistic model, instead measured pressure drop data from the experiments were directly used. 398 different experiments were performed using different liquid and gas flow rates and corresponding pressure gradient values are captured. The results indicate that the computational intelligence models performed better than the mechanistic model in predicting the pressure drop. Among the computational intelligence models, Support Vector Regression (SVR) provided the closest results to the actual experimental data due to its relative better tolerance to higher error terms. Moreover, all models provided acceptable error performance; as such they can be used as good pressure drop estimation models for this problem.

ACKNOWLEDGMENTS

This study was funded through TUBITAK (The Scientific and Technological Research Council of Turkey) Project 108M106.

REFERENCES

1.   Alizadehdakhel A, Rahimi M, Sanjari J, Alsairafi AA. CFD and artificial neural network modeling of two-phase flow pressure drop. Int Commun Heat Mass Transfer 2009;36(8):850–6.

2.   Barnea D. A unified model for predicting flow pattern transitions for the whole range of pipe inclinations. Int J Multiphase Flow 1987;13(1):1.

3. Beggs HD, Brill JP. A study of two-phase flow in inclined pipes. J Petrol Technol 1973:607. Trans., AIME, 255.

4. Bendiksen KH, Malnes D, Moe R, Nuland S. The dynamic two-fluid model OLGA: theory and application. SPE Prod Eng May 1991;6:171–80.

5. Belt R, Djoric B, Kalali S, Duret E, Larrey D. Comparison of commercial multiphase flow simulators with experimental and field databases. In: Presented at BHR Group's multiphase production technology conference in Cannes; 2011.

6. Caetano EF, Shoham O, Brill JP. Upward vertical two-phase flow through an annulus. Part – I: Single phase friction factor, Taylor Bubble velocity and flow pattern prediction. ASME J Energy Resour Technol 1992;114:1–13.

7. Drucker H, Burges CJC, Kaufman L, Smola A, Vapnik V. Support vector regression machines. In: Mozer MC, Jordan MI, Petsche T, editors. Advances in Neural Information Processing Systems, vol. 9. Cambridge, MA: MIT Press; 1997. p. 155–61.

8. Dukler AE, Wickes III M, Cleveland RG. Frictional pressure drop in two-phase flow: an approach through similarity analysis. AIChE J 1964;10(1):44.

9. Duns Jr. H, Ros NCJ. Vertical flow of gas and liquid mixtures in wells. In: Proc, sixth world petroleum congress, Frankfurt-am-Main, Germany; 1963.

10. Erge O, Ozbayoglu ME, Miska S, Yu M, Takach N, Saasen A, et al. Effect of drillstring deflection and rotary speed on annular frictional pressure losses. J Energy Res Technol 2014. http://dx.doi.org/10.1115/1.4027565.

11. Erge O, Ozbayoglu ME, Miska S, Yu M, Takach N, Saasen A, et al. The effects of drillstring eccentricity, rotation and buckling configurations on annular frictional pressure losses while circulating yield power law fluids. Paper SPE 167950 presented at the IADC/SPE drilling conference & exhibition, 4–6 March, Fort Worth, Texas; 2014. http://dx.doi.org/10.2118/167950-MS.

12. Osgouei Ettehadi, Liew W, Ozbayoglu ME. Calculations of

equivalent circulating density in underbalanced drilling operations. In: 6th International petroleum technology conference, Beijing, China; March 26–28, 2013.

13. Hasan AR, Kabir CS. Two-phase flow in vertical and inclined annuli. Int J Multiphase Flow 1992;18:279–93.

14. Haykin S. Neural networks: a comprehensive foundation. 2nd ed. Upper Saddle River, NJ: Prentice Hall; 1999.

15. Karimi Vajargah A, Miska S, Yu M, Majidi R, Ozbayoglu ME. Feasibility study of applying intelligent drill pipe in early detection of gas influx during conventional drilling. SPE/IADC 163445, SPE/IADC drilling conference & exhibition, Amsterdam, The Netherlands; 5–7 March, 2013.

16. Karimi Vajargah A, Miska S, Yu M, Ozbayoglu ME, Majidi R. Taking the proper action to gas influx during constant bottom-hole pressure technique of managed pressure drilling, OTC 24189. In: 2013 Offshore Technology Conference, Houston TX, USA; 6–9 May, 2013.

17. Lage ACVM, Rommetveit R, Time RW. An experimental and theoretical study of two-phase flow in horizontal or slightly deviated fully eccentric annuli, SPE 62793, IADC/SPE Asia Pacific Drilling Technology, Kuala Lumpur, Malaysia; 11–13 September, 2000.

18. Lockhart RW, Martinelli RC. Proposed correlation of data for isothermal twophase, two-component flow in pipes. Chem Eng Progress 1949;45(1):39–48.

19. Mattera D, Haykin S. Support vector machines for dynamic reconstruction of a chaotic system. In: Scholkopf B, Burges CJC, Smola AJ, editors. Advances in Kernel methods—support vector learning. Cambridge, MA: MIT Press; 1999. p. 211–42.

20. Muller K-R, Smola A, Ratsch G, Scholkopf B, Kohlmorgen J, Vapnik V. Predicting time series with support vector machines. In: Gerstner W, Germond A, Hasler M, Nicoud J-D, editors. Artificial neural networks ICANN'97, vol. 1327. Berlin: Springer Lecture Notes in Computer Science; 1997. p. 999–1004.

21. Osgouei RE, Ozbayoglu ME, Ozbayoglu AM. A mechanistic model to characterize the two phase drilling fluid flow through inclined eccentric annular geometry. In: SPE oil and gas India conference and exhibition, Mumbai, India; 28–30 March, 2012.

22. Osman E-SA. Artificial neural network models for identifying flow regimes and predicting liquid holdup in horizontal multiphase flow, SPEPF; February 2004. p. 33.

23. Ozbayoglu EM, Ozbayoglu MA. Estimating flow patterns and frictional pressure losses of two-phase fluids in horizontal wellbores using artificial neural networks. Pet Sci Technol 2009;27(2):135–49.

24. Ozbayoglu ME, Omurlu C. Two-phase flow through fully eccentric horizontal annuli: a mechanistic approach. SPE 107076, presented at SPE/ICOTA coiled tubing and well intervention conference and exhibition, The Woodlans, Texas; 20–21 March, 2007.

25. Petalas N, Aziz K. A mechanistic model for multiphase flow in pipes. JCPT 2000;39(6):43–55.

26. Russell S, Norvig P. Artificial Intelligence: a modern approach. 3rd ed. Pearson Prentice Hall; 2010.

27. Sadatomi M, Sato Y, Saruwatari S. Two-phase flow in vertical noncircular channels. Int J Multiphase Flow 1982;8:641–55.

28. Smola AJ, Schölkopf B. Stat Comput 2004;14:199–222.

29. Stitson M, Gammerman A, Vapnik V, Vovk V, Watkins C, Weston J. Support vector regression with ANOVA decomposition kernels. In: Scholkopf B, Burges CJC, Smola AJ, editors. Advances in Kernel methods—support vector learning. Cambridge, MA: MIT Press; 1999. p. 285–92.

30. Sunthankar AA. Study of the flow of aerated drilling fluids in annulus under ambient temperature and pressure conditions. Ms. Thesis, the University of Tulsa 2000.

31. Taitel Y, Dukler AE. A model for predicting flow regime transition in horizontal and near horizontal gas–liquid flow. AIChE J 1976;22(1):47.

32.   Ternyik J, Bilgesu HI, Mohaghegh S. Virtual measurements in pipes. Part 2: Liquid holdup and flow pattern correlations. Eastern regional meeting, Morgantown, West Virginia, USA; September 1995.

33.   Theodoridis S, Koutroumbas K. Pattern recognition. 4th ed. Academic Press; 2009. ISBN: 978-1-59749-272-0.

34.   Timung S. Prediction of flow pattern of gas–liquid flow through circular microchannel using probabilistic neural network. Appl Soft Comput April 2013;13(4):1674–85.

35.   Vapnik V, Chervonenkis A. A note on one class of perceptrons. Automat Remote Contr 1964;25.

36.   Vapnik V. The nature of statistical learning theory. New York: Springer; 1995.

37.   Wallis GB. One-dimensional two-phase flow. New York City: McGraw-Hill Book Co., Inc.; 1969.

38.   Xiao JJ, Shoham O, Brill JP. A comprehensive mechanistic model for two-phase flow in pipelines. Paper SPE 20631 presented at the 1990 SPE annual technical conference and exhibition, New Orleans; 23–26 September, 1990.

39.   Zhao B. Modeling pressure drop coefficient for cyclone separators: a support vector machine approach. Chem Eng Sci 2009;64(19):4131–6.

Chapter **8**

Experimental Investigation of Use of Horizontal Wells in Waterflooding

N. Hadia[a], L. Chaudhari[a], Sushanta K. Mitra[a], M. Vinjamur[b], and R. Singh[c]

[a]Department of Mechanical Engineering, IITB-ONGC Joint Research Centre, Indian Institute of Technology Bombay, Mumbai, 400 076 India

[b]Department of Chemical Engineering, Indian Institute of Technology Bombay, Mumbai, 400 076 India

[c]Institute of Reservoir Studies, ONGC, Ahmedabad, India

ABSTRACT

The present experimental study is conducted to investigate the effect of horizontal production well in waterflooding when conventional vertical production well stops oil production. Laboratory experiments have been performed on three-dimensional sand pack models. In the waterflooding experiments, the horizontal production well is turned on after the complete extraction of oil from the vertical production well. The experimental results show that horizontal well increases the ultimate recovery by waterflooding significantly when oil production stops from the conventional vertical production well. Moreover, the position of the horizontal well either at top or bottom of the sand pack has negligible effect on ultimate oil recovery for the sand packs used.

INTRODUCTION

During oil exploration, it is commonly found that after the primary oil recovery, still large quantity of unexplored oil remains in place. Various recovery processes viz., waterflooding, gas injection, thermal recovery, chemical flooding, etc. are then applied to recover this oil from the reservoir. The main objective of these recovery mechanisms is to increase the ultimate production from reservoirs. Waterflooding, in which injected water provides a driving mechanism to recover the oil, is simple, inexpensive secondary recovery process and is being widely used for the oil recovery after primary oil recovery process. Applications of horizontal wells in waterflooding projects as both injectors and producers continue to grow. Horizontal wells have been used in waterflood and in polymer flood applications to improve sweep efficiency (Joshi, 1991). The advantage of horizontal wells in waterflood/enhanced oil recovery (EOR) applications are enhanced injectivity and productivity. Another main advantage is their ability to reduce the number of vertical injection and production wells without sacrificing injectivity or productivity (Joshi, 1999).

In spite of a tremendous increase in the use of horizontal wells for reservoir exploitation, most of the applications are found for EOR methods viz., thermal recovery (Joshi, 1986, Bagci and Gumrah, 1992 and Guanghul et al., 1995), chemical and polymer flooding (Bagci and Hodaie, 2003 and Bagci, 2004), steam-CO_2

drive experiments (Gumrah and Bagci, 1997), immiscible CO_2 and WAG injection (Erdal and Bagci, 2000), etc. Most of these studies have shown that the horizontal wells improve the ultimate oil recovery over that of conventional vertical wells during EOR.

A significant number of numerical simulation studies are available on use of horizontal well for waterflooding projects. Pieters and Al-Khalifa (1991), using three-dimensional reservoir simulation model, investigated the use of horizontal and vertical wells in waterflooding for a layered heterogeneous carbonate reservoir. They showed that horizontal and vertical wells recovered the same amount ofoil in tight reservoirs provided the vertical well penetrates the entire reservoir.

Dykstra and Dickinson (1992) calculated the gravity drainage oil recovery from vertical and horizontal wells. They stated that for flat formations (no gravity effect), at thickness less than 0.85 times the well spacing, a horizontal well produces better than the vertical well whereas, at formation thickness greater than this, a vertical well performs better. Also, for flat formations, the formation thickness affects the ratio of horizontal/vertical well flow rates. But for dipping formations, formation thickness has no effect on the ratio of horizontal/vertical well flow rates. Taber and Seright (1992) showed the advantages of using horizontal wells in combination with vertical wells in waterflooding. They showed that the use of horizontal well can increase the areal sweep efficiency by 25% to 40%. They also suggested that horizontal wells are more advantageous in thin formations than thick formations for waterflooding projects.

Using two-dimensional reservoir simulation studies, Joshi et al. (1993) showed that horizontal wells, used as producers or injectors do not provide a significant increase in areal sweep efficiency over vertical wells. However, horizontal wells have higher productivity as producers and higher injectivity as injectors. Moreover, the reservoirs with high permeability, horizontal wells may not provide a significant advantage. Two- and three-dimensional simulation studies performed by Ferreira et al. (1996) showed that vertical to horizontal permeability ratio, injection and production rate, and reservoir thickness have little effect on waterflood oil recovery for a particular mobility ratio. They observed that waterflood performance is better with horizontal well as compared to conventional vertical well. They developed a correlation that expresses the volumetric sweep efficiency as a function of mobility ratio and is useful to predict the waterflood recovery.

Gharbi et al. (1996), using three dimensional chemical flood simulator, investigated the performance of immiscible displacement with horizontal and vertical wells in heterogeneous reservoirs. They studied the sensitivity of the displacement performance to the horizontal well length and the ratio of horizontal to vertical permeability using various well combinations. They showed that the degree and structure of the heterogeneity of the reservoir have a significant effect on the efficiency of immiscible displacement with horizontal wells. Long horizontal wells in highly heterogeneous reservoir do not necessary guarantee improved oil recovery. In subsequent work, Gharbi et al. (1997) showed that the performance of enhanced oil recovery processes with horizontal wells is strongly affected by the permeability variation and the spatial correlation of the reservoir heterogeneity.

Popa and Clipea (1998) studied the effect of pressure drop along the horizontal well section and geometry of the well pattern on waterflood sweep efficiency. They showed that it is preferable to use vertical wells as injectors and horizontal wells as producers rather than the use of horizontal wells as injectors and verticals well as producers. Algharaib and Ertekin (1999) studied the effect of various waterflooding fluid parameters together with some operational design parameters. The numerical analysis showed that the combination, in which one horizontal and one vertical well are utilized, performs similar to the combination of two horizontal wells. Popa et al. (2002) analyzed the overall efficiency of a waterflooding process that is influenced by well pattern using horizontal/multilateral injectors and producers in different configurations. They showed that main parameters, such as breakthrough time, oil recovery at breakthrough, sweep efficiency, injection-production pressure, etc. are strongly affected by the type of configuration considered. Recently, Algharaib and Gharbi (2005) investigated the performance of non-conventional wells in water flooding projects under different operating/reservoir conditions using numerical simulation techniques. Their results show that the well pattern used for waterflooding has a significant effect on the displacement performance of non-conventional wells. Moreover, long horizontal/multilateral wells do not automatically guarantee improved oil recovery. A very limited experimental studies have been conducted to investigate the effect of horizontal well on waterflooding oil recovery. Shirif et al. (2003) with the help of experimental study, examined the effect of vertical and horizontal injection and production well combinations and found that the use of horizontal wells showed slightly better oil recovery over vertical wells in a waterflood of reservoirs under bottom water conditions.

From the existing literature, it can be concluded that horizontal wells are advantageous in EOR over conventional vertical wells. The main objective of this paper is to investigate, by experimental studies, the performance of horizontal production well in the later stage of waterflooding i.e., when oil recovery ceases from conventional vertical injection–vertical production (VI–VP) well configuration. Waterflooding experiments have been performed on three-dimensional sand pack models. Initially, each flooding experiment is performed with VI–VP well pattern till oil recovery stops and then horizontal production well is opened and observed for additional oil recovery, if any. Moreover, horizontal production wells are placed in the top and bottom part of the reservoir to investigate the effect of placement of horizontal well on oil recovery.

EXPERIMENTS

The schematic of experimental apparatus is shown in Fig. 1. The apparatus consists of a dual cylinder precision syringe pump for injection, a three-dimensional core holder (Mitra et al., 2005) to hold sand pack, and a differential pressure transmitter to measure the pressure drop across the sand pack.

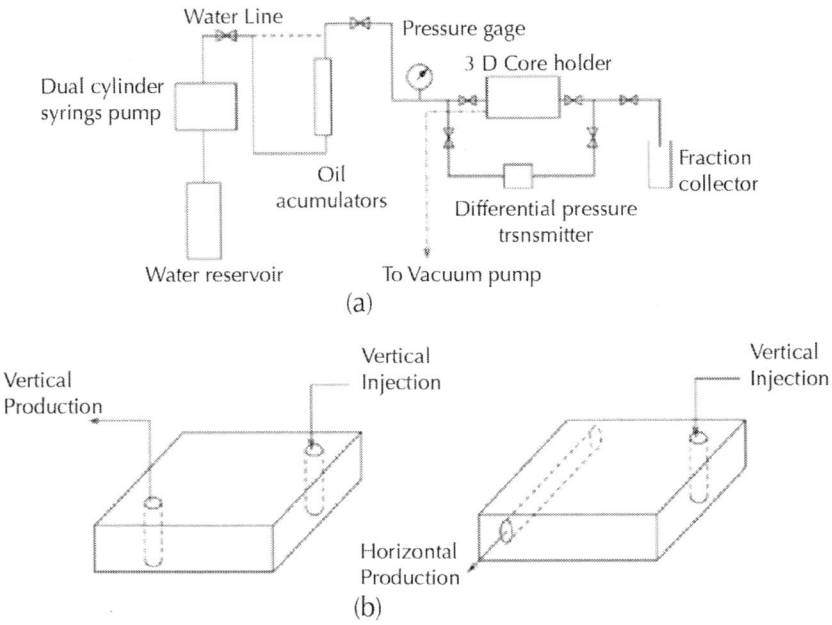

Figure 1: (a) Schematic of experimental apparatus (dotted line shows by-pass for waterflooding). (b) Well configuration in three-dimensional sand pack.

The inside dimensions of a specially designed three-dimensional core holder are 31 cm in length, 31 cm in width and 12 cm in depth. The location of vertical injection and production wells are shown in Fig. 2. Horizontal production wells are used at top and bottom of the sand pack as shown in Fig. 3. The wells are made of stainless steel tubes of 0.6 cm inner diameter fully perforated along the length with a perforation density of 28 holes per square inch. Each perforation is of 0.3 cm in diameter. The wells are covered with 150 mesh size metal screen to prevent the entry of sand into the wells.

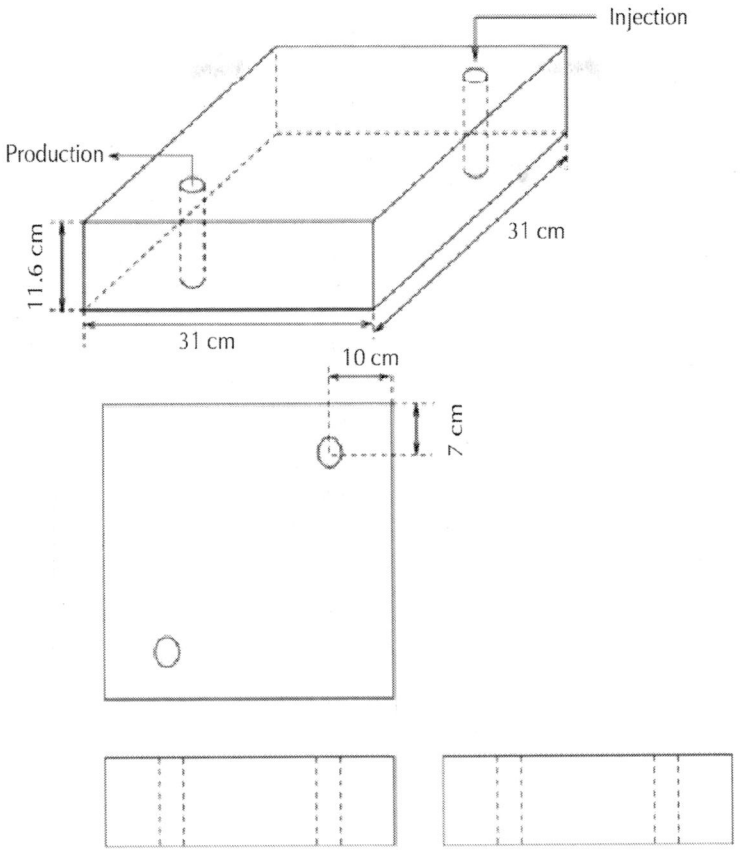

Figure 2: Vertical injection and production well locations in three-dimensional sand pack.

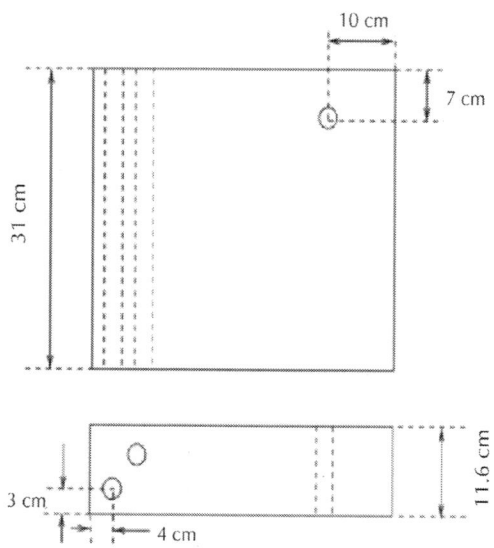

Figure 3: Top and bottom horizontal production well locations in three-dimensional sand pack.

Clean and washed sand of mesh size between ASTM 30 and 50 is used as packing material. Water and paraffin oil are used as displacing and displaced fluids, respectively. The properties of paraffin oil are provided in Table 1. The viscosity of oil is measured with rheometer with an accuracy of ± 1% of measured value of viscosity.

Table 1: Properties of paraffin oil

Fluid	Property	Value
Paraffin oil	Specific Gravity @ 38 °C	0.8545
	Viscosity @ 25 °C	130 mPa.s
Water	Viscosity @ 25 °C	0.97 mPa.s

The clean and washed sand is packed in the coreholder and wells are placed at desired locations. The sand pack is covered with Teflon sheet and steel plate. The steel plate is then fastened with main core holder body by applying uniform torque by torque wrench to each bolt. The sand pack is then tested for leakage with soap bubble test using high pressure nitrogen up to 2 bar pressure. When pressure is maintained for half an hour, it is assumed that the pack is leak-proof. The pack is then connected to a vacuum pump for evacuation (740 mm Hg of vacuum is achieved). After completing the evacuation procedure, the saturation process is carried out with water to determine the pore volume (PV) of the sand pack.

In water saturation process, the water is allowed to flow by gravity into the sand pack. Pore volume is then determined by the amount of water absorbed by the pack. The average absolute permeability of the pack is then determined by flow of water at a constant rate and measuring the pressure drop between two ends of the sand pack. To ensure the injected fluid entered the pack over its entire cross-section, the injection is carried out simultaneously at different locations on the face of sand pack using common manifolds of the core holder. Similarly, the effluent is also collected using common manifold at the outlet and the pressure drop is measured between inlet and outlet manifolds.

The properties of two different sand pack samples viz., SP1 and SP2, are provided in Table 2. Paraffin oil is then injected to displace the water to irreducible water saturation. The condition of irreducible water saturation is ensured when no more water is observed in the effluent. After the model is prepared for the displacement experiments, waterflooding is carried out using different well configurations. The constant injection rate of 300 ml/h is used for all the experiments. All waterflooding experiments were performed at constant room temperature of 25 ± 1 °C and atmospheric pressure.

Table 2: Properties of three-dimensional sand packs

Sample	Porosity	Absolute permeability (mD)
SP1	41.1	3125
SP2	44.5	3400

After completion of waterflood with a particular well configuration, the same pack is re-flooded with oil to displace the water to irreducible water saturation and waterflooding experiments are performed for other well configurations. The generation of one set of recovery curve requires an experimentation time of 1 week including the regeneration (re-saturation with oil) time of 3 d.

For sand pack SP1, after re-saturating it with oil, a settling time (ageing) of 24 h is provided before start of waterflood. Moreover, the horizontal well is opened after 12 h of completion of vertical injection-vertical production (VI–VP) configuration. For sample SP2, however, no such settling time is allowed. Waterflooding is carried out soon after the re-saturation with oil is completed and the horizontal well is opened soon after the completion of VI–VP configuration.

RESULTS AND DISCUSSION

Displacement experiments are conducted on three dimensional sand packs with different well configurations to investigate the effectiveness of horizontal well over conventional vertical wells in waterflooding. Experimental conditions together with final and breakthrough recovery are given in Table 3. In reference to Table 3, Run-1 was first conducted and the sand pack is regenerated by flooding with oil. Then Run-2 is performed and it is followed by Run-3.

Table 3: Experimental conditions for three-dimensional sand packs

Sample	Run	S_o	S_{wi}	Breakthrough	VI–VP	VI–HP
		% PV	% PV	Recovery, % OOIP	Recovery, % OOIP	Recovery[a], % OOIP
SP1	1	84.1	15.9	19.2	59.0	6.1
	2	83.7	16.3	21.1	59.4	5.3

SP2	1	77.1	22.3	13.7	56.1	6.1
	2	80.2	19.8	12.6	51.2	6.1
	3	82.2	17.8	88.9	47.8	–

S_o = residual oil saturation.

S_{wi} = irreducible water saturation.

OOIP = original oil in place.

[a]Additional recovery after VI–VP configuration.

Four waterflood experiments are conducted with two different sand packs. Each sand pack has been reused after completion of one set of experiment. In all experiments, initially waterflooding is performed with vertical injection-vertical production (VI–VP) configuration. When recovery ceases from the vertical production well, the horizontal well is opened and incremental recovery is observed.

Fig. 4 is the recovery curve for Run-1 of sample SP1. It is observed that VI–VP produces 59% of original oil inplace (OOIP). After about 2.4 PV of injection, oil production ceased from vertical well; then the horizontal well is opened. After 1 PV of additional injection, horizontal well recovered 6.1% of OOIP which is equivalent to 14.8% of residual oil saturation left in the sand pack after VI–VP configuration. Similarly, in case of Run-2, VI–VP produces 59.4% of OOIP after 2.4 PV of injection and ceases to produce anymore as shown in Fig. 5. Opening of the horizontal well then after produces 5.3% of OOIP for an additional 1 PV of water injection. It is to be noted that for Run-1, the horizontal well is placed at top (HWT), whereas for Run-2, it is placed at the bottom (HWB) of the sand pack. However, such placement is found to have negligible effect on the ultimate oil recovery.

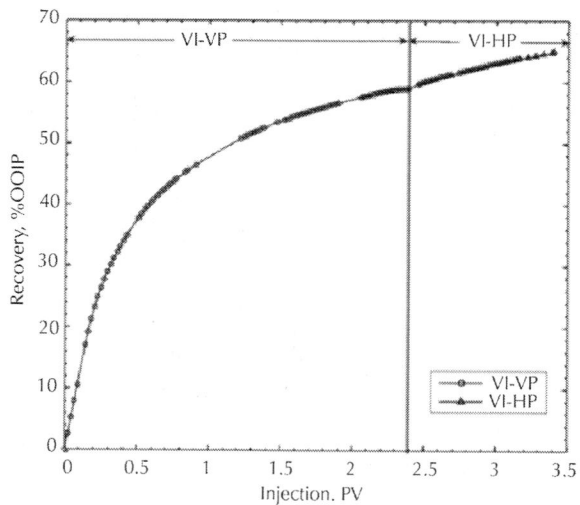

Figure 4: Recovery curve when VI–VP used first and is followed by VI–HP for Run-1 of sample SP1. Horizontal well placed at top (HWT).

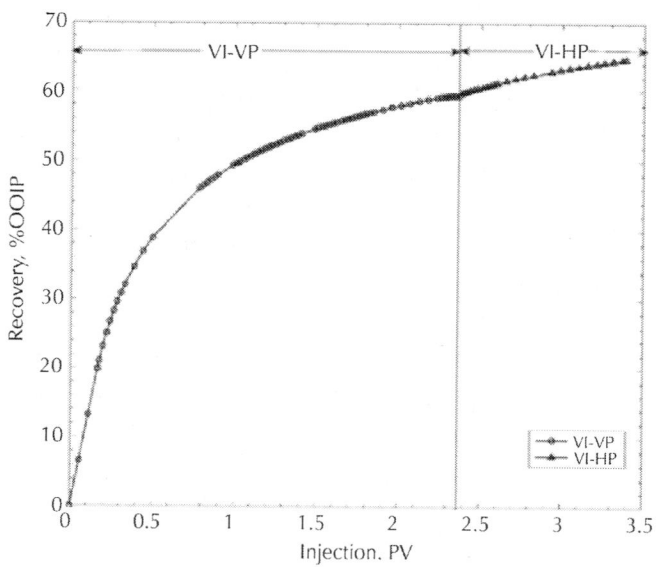

Figure 5: Recovery curve when VI–VP is used first and is followed by VI–HP for Run-2 of sample SP1. Horizontal well placed at bottom (HWB).

The oil recovery plots for both Run-1 and Run-2 are overlayed on top of each other as shown in Fig. 6. It is observed that for both the runs, the ultimate recovery obtained with VI–VP configuration is same, which indicates that the sand pack can be reused. Moreover, in both the runs, horizontal production well provides nearly the same incremental recovery after production stops from the vertical well. This additional recovery by horizontal well can be attributed to the large reservoir contact area of horizontal well compared to the vertical well. The reservoir area which remains unswept when vertical well is used would be swept when horizontal well is used and this results in an increase in the recovery. It is to be noted that the settling time (ageing) of 24 h is allowed each regeneration of the sand pack.

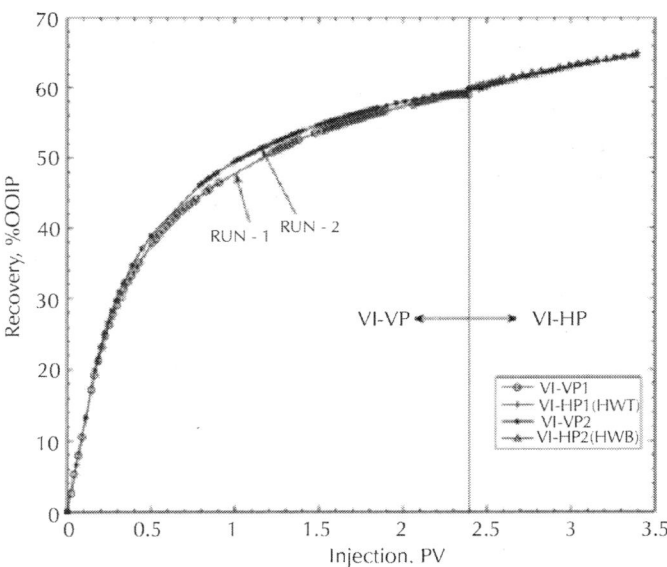

Figure 6: Recovery comparison when VI–VP used first and is followed by VI–HP for Run-1 and Run-2 of sample SP1.

As mentioned earlier, additional waterflooding performed on another sample SP2, but under different ageing condition. The effect of horizontal well, for Run-1 and Run-2, in waterflooding for sand pack sample SP2 is shown in Fig. 7 and Fig. 8, respectively and these two curves are overlayed in Fig. 9. It is to be noted that for Run-1, the horizontal well is placed at bottom (HWB), whereas for Run-2, it is placed at the top (HWT). It can be observed that oil recovery falls about by 5% in successive runs, i.e., Run-1 and Run-2, for VI–VP configuration. However, the incremental oil recovery of 6% with horizontal production well is same for both runs. For Run-3, VI–VP configuration produced 47.8% of OOIP after 2.6 PV of water injection as shown in Fig. 9. The difference in oil recovery for VI–VP configuration can be attributed to the immediate starting of waterflood after connate water saturation is established. After each oil saturation, the sand pack tends to become more oil-wet. Hence, the oil saturation increases after each oil resaturation. The comparison of oil saturation values for sample SP2 from Table 3 clearly shows that as number of runs increases, the oil saturation increases and sand pack attains more wettability to oil. This results in change in the relative permeability curves and hence reduction in ultimate oil recovery (Grattoni et al., 2000). It is observed, however, that the positions of the horizontal well in the sand pack have negligible effect on ultimate oil recovery.

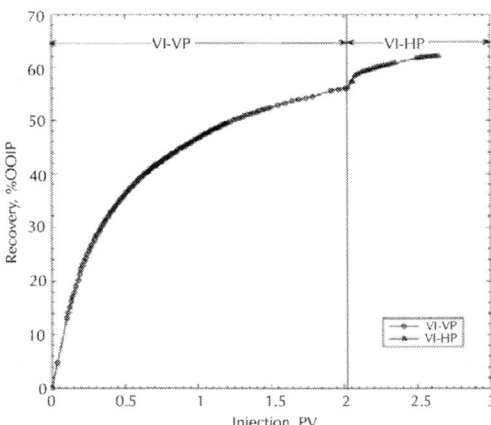

Figure 7: Recovery curve when VI–VP used first and is followed by VI–HP for Run-1 of sample SP2. Horizontal well placed at bottom (HWB).

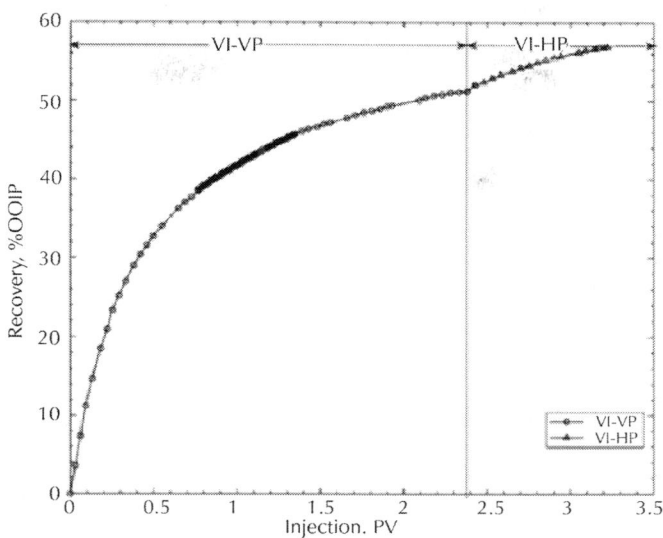

Figure 8: Recovery curve when VI–VP used first and is followed by VI–HP for Run-2 of sample SP2. Horizontal well placed at top (HWT).

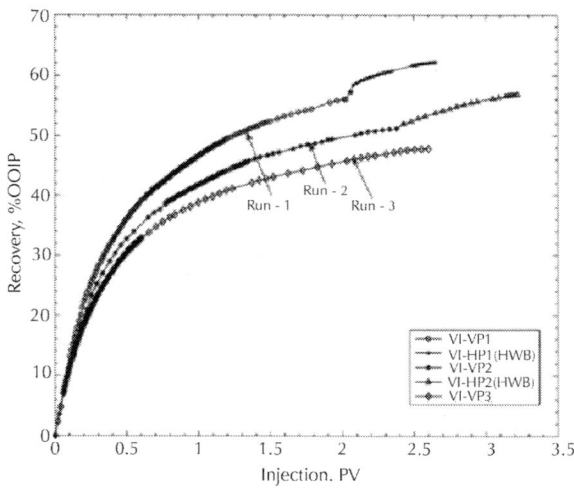

Figure 9: Recovery comparison when VI–VP used first and is followed by VI–HP for Run-1, Run-2, and Run-3 of sample SP2. The oil saturation

level and recovery fall every-time the core is regenerated and not kept for ageing. Sequence of operation: Run-1 $\rightarrow$ Run-2 $\rightarrow$ Run-3.

CONCLUSIONS

The performance of horizontal production well, in later stage of waterflooding, on oil recovery has been investigated here. Experimental studies have been conducted using three-dimensional sand pack models. From present experimental investigation, it can be concluded that the horizontal production well increases the ultimate oil recovery in waterflooding significantly, when used after the oil production stops from conventional vertical injection–vertical production well configurations. Also, the placement of the horizontal production well has negligible effect on the ultimate oil recovery for the sand packs considered. Moreover, when settling time is allowed after re-saturation of the pack with oil, the pack can be reused for experiments.

REFERENCES

1.   Algharaib, M., Ertekin, T., 1999. The efficiency of horizontal and vertical well patterns in waterflooding: a numerical study. Paper SPE 52196 Presented at the 1999 SPE Mid-Continent Operations Symposium held in Oklahoma City, Oklahoma, March 28–31.

2.   Algharaib, M., Gharbi, R.B.C., 2005. A comparative analysis of waterflooding projects using horizontal wells. Paper SPE 93743 Presented at the 2005 Middle East Oil Show (MEOS) held in Behrain, March 12–15.

3.   Bagci, S., 2004. 3-D scaled model studies of alkaline flooding using horizontal well. Energy Sources 26, 783–793.

4.   Bagci, S., Gumrah, F., 1992. An examination of steam injection process in horizontal and vertical wells for heavy oil recovery. J. Pet. Sci. Eng. 8, 59–72.

5.   Bagci, S., Hodaie, H., 2003. An investigation of polymer flooding in limestone reservoirs with bottom water zone. Energy Sources 25, 253–264.

6.   Dykstra, H., Dickinson, W., 1992. Oil recovery by gravity drainage into horizontal wells compared with recovery from vertical wells. SPE Form. Eval. 255–260 (September).

7.   Erdal, T., Bagci, S., 2000. Scaled 3-D model studies of immiscible CO_2 flooding using horizontal wells. J. Pet. Sci. Eng. 26, 67–81.

8.   Ferreira, H., Mamora, D.D., Startzman, R.A., 1996. Simulation studies of waterflood performance with horizontal well. Paper SPE 35208 Presented at Permian Basin Oil and Gas Recovery Conference held in Midland, Texas, March 27–29.

9.   Gharbi, R.B., Peters, E.J., Afzal, N., 1996. Effect of heterogeneity on the performance of immiscible displacement with horizontal well. Paper SPE/DOE 35441 Presented at the 1996 SPE/DOE Tenth Symposium on IOR held in Tulsa, OK, April 21–24.

10. Gharbi, R.B., Peters, E.J., Elkamel, A., Afzal, N., 1997. Effect of heterogeneity on the performance of EOR processes with horizontal wells. Paper Presented at 6th International Symposium on Evaluation of Reservoir Wettability and Its effect on Oil Recovery held at Socorro, New Mexico, USA, September 27–28.

11. Grattoni, C.A., Jing, X., Zimmerman, R.W., 2000. Wettability alteration by ageing of a gel placed within a porous medium. Paper SPE 38320 Presented at the 1997 SPE SPE Western Regional Meeting held in Long Beach, California, June 25–27.

12. Guanghul, Z., Zhang, R., Shen, D., Pu, H., 1995. Horizontal well application in a high viscous oil reservoirs. Paper SPE 30281 Presented at the International Heavy Oil Symposium held in Calgary, Alberta, Canada, June 19–21.

13. Gumrah, F., Bagci, S., 1997. Steam-CO_2 drive experiments using horizontal and vertical wells. J. Pet. Sci. Eng. 18, 113–129.

14. Joshi, S.D., 1986. A laboratory study of thermal oil recovery using horizontal wells. Paper SPE/DOE 14916 Presented at Fifth Symposium on EOR of SPE and DOE held in Tulsa, OK, April 20–23.

15. Joshi, S.D., 1991. Horizontal Well Technology. PennWell Publishing Company, Tulsa, OK, p. 12.

16. Joshi, S.D., 1999. Special report: horizontal and unbalanced drilling. The American Oil and Gas Reporter, pp. 60–63.

17. Joshi, S.D., Ding, W., Hall, K., 1993. A simulation study of waterflooding using combinations of horizontal and vertical wells. SVIP 005 Presented at Tenth Petroleum Engineering Conference of the SPE of Venezuela, Puerto La Cruz, Venezuela, October 20–25.

18. Mitra, S.K., Vinjamur, M., Singh, R., Maurya, C.R., 2005. ThreeDimensional Core Holder. Filed Indian Patent Application No. 1463/MUM/2005.

19. Pieters, D.A., Al-Khalifa, A.J., 1991. Horizontal well performance in a layered carbonate reservoir. Paper SPE 21865 Presented at the Rocky Mountain Regional Meeting and Low-Permeability Reservoirs Symposium held in Denver, Colorado, April 15–17.

20. Popa, C.G., Clipea, M., 1998. Improved waterflooding efficiency by horizontal wells. Paper SPE 50400 Presented at the 1998 SPE International Conference on Horizontal Well Technology held in Calgary, Alberta, Canada, November 1–4.

21. Popa, C.G., Romania, P., Turta, T.T., 2002. Waterflooding by horizontal injectors and producers. Paper SPE/Petroleum Society of CIM/CHOA 78989 Presented at the 2002 SPE International Thermal Operations and Heavy Oil Symposium and International Horizontal Well Technology Conference held in Calgary, Alberta, Canada, December 4–7.

22. Shirif, E., ElKaddifi, K., Hromek, J.J., 2003. Waterflood performance under bottom water conditions: experimental approach. SPE Reservoir Engineering and Evaluation, pp. 28–38. February.

23. Taber, J.J., Seright, R., 1992. Horizontal injection and production wells for EOR or waterflooding. Paper SPE 23952 Presented at the 1992 SPE Permian Basin Oil and Gas Recovery Conference held in Midland, Texas, March 18–20.

Numerical Simulation of Electricity Generation Potential from Fractured Granite Reservoir through a Single Horizontal Well at Yangbajing Geothermal Field

Yu-Chao Zeng, Neng-You Wu, Zheng Su, and
Jian Hu

Key Laboratory of Renewable Energy, Guangzhou Institute of Energy
Conversion, Chinese Academy of Sciences, No.2 Nengyuan Road,
Guangzhou 510640, China

ABSTRACT

Development of deep high-temperature heat reservoir at Yangbajing geothermal field has very important significance for capacity expanding and sustaining of the ground power plant. The geological exploration found that there is a fractured granite heat reservoir with an average temperature of 248°C at depth of 950~1350 m in well ZK4001 in the north of the geothermal field. In this work, a simplified conceptual model of the 950~1350 m reservoir is established based on all the existing geological data, and the electricity generation potential from this fractured granite reservoir by geothermal water mining through a single horizontal well is numerically simulated. The results indicate that the single horizontal well system attains an electric power of 3.23~3.48 MW and an energy efficiency of about 50.00~17.16 during 20 years under reference conditions. The main parameters that affect the heat extraction and electricity generation are reservoir porosity, permeability and water production rate. Higher porosity or higher permeability or higher water production rate will be favorable for improving the electricity generation performance, under the precondition of not arousing vaporization and precipitation in the liquid water saturated reservoir.

INTRODUCTION

Background

HDR (hot dry rock) geothermal energy is heat energy stored in subsurface hot and low permeable crystalline rocks, which normally locates at depth within 3~10 km [1]. Total HDR geothermal resource within 10 km depth all over the world amounts to about 40~400M EJ (1 EJ = 10^{18} J), approximately 100~1000 times the quantity of fossil energy [2] and [3]. In America, total amount of high-grade HDR resource with geothermal gradient higher than 45

°C/km within 10 km depth is about 6.5 × 10⁵ EJ, far more than that of world fossil energy [3]. Furthermore, the HDR resource is well concentrated and stable, very suitable for generating base-load electric power [1], [2] and [3]. Heat mining system by drilling wells, hydrofracturing rocks and circulating water through aperture network within the fractured reservoir is termed as EGS (enhanced geothermal system), and normally the produced heat is used for electricity generation [1]. EGS will provide about 100,000 MW electric power for America by 2050, occupying about 10% of total electricity generating capacity in the USA [1].

The world EGS tests began in 1974, when the Los Alamos National Laboratory started to conduct the heat mining experiment from HDR resource at Fenton Hill. This pioneering field test lasted from 1974 to 1993 and firstly proved that the heat in the HDR can be successfully exploited [1]. Inspired by the Fenton Hill test, U.K. conducted Rosemanowes EGS test from 1976 to 1992, Germany conducted Falkenberg EGS test in 1976–1985 and Urach EGS test in 1978–1981, France conducted Le Mayet test in 1977–1980, and Japan conducted Ogachi test in 1982–2004 and Hijiori test in 1988–2002 [1] and [4]. These field tests firmly proved that the drill of deep well, simulation of pre-existing sealed fractures, and interwell connection can be successfully conducted, the stable water circulation can be well sustained, and the circulated water can be heated to the target temperature [1]. The main obstacles to develop EGSs are that the drilling cost is too expensive, the water circulation rate is too low and the water flow impedance is too high [1], [4] and [5]. The present on-going EGS field tests include the second phase of Le Mayet EGS test since 1981 and Soultz EGS test since 1986 in France, the second phase of Urach EGS test since 1990 in Germany, Hunter Valley test since 1999 and Cooper Basin test since 2002 in Australia and Coso test and Desert Peak test since 2002 in America [1] and [4]. At Soultz test field, some natural water was found in the fractures and faults in the rocks, thus the concept of "hot dry rock" was expanded to "HWR" (hot wet rock), which normally locates at margin of traditional hydrothermal system [5]. The purely waterless hot dry rock system, the hot wet rock system and the hydrothermal system constitute a spectrum of geothermal resource type [5]. Generalized definition of hot dry rock system mainly emphasizes the reservoir is comprised of hot fractured rocks, and whether there is pre-existing water is not the key point, so in this study we include the naturally hot fractured granite reservoir at Yangbajing geothermal field into the hot dry rock resource type [1] and [5].

The research on HDR in China originates from electricity generation test using HDR at Fangshan District in Beijing conducted by China Seismological Bureau and Japan Chuo University from 1993 to 1995 [6]. Since 2000 Zhao has led a research group to conduct systematical studies on HDR resource exploiting in China [7] and [8]. In 2007 China Geothermal Professional Committee and Australia Petratherm Company launched a cooperative study for two years on China resource potential of EGS [9]. Influenced by collision and subduction from Indian Ocean plate, Philippine sea plate and Pacific plate to China's continental plate, huge potential high-temperature geothermal resource zone are formed at south Tibet, West Yunnan province, north Hainan province, Changbai Mountain and Wudalianchi district in China [10]. Total HDR resource reserve in China within 3~10 km depth amounts to 20.90M EJ; if we take 2% as the recoverable fraction, the recoverable HDR resource amounts to 168 times the quality of traditional hydrothermal resource or 4400 times total annual energy consumption in 2010 in China [11]. The most promising areas for HDR exploiting test are Yangbajing and Yangyi district in Dangxiong County in Tibet, Tengchong district in Yunnan Province and north Hainan Province [7], [8], [10] and [11]. Specially, the Yangbajing geothermal power plant has an installed capacity of 27.30 MW now from shallow hydrothermal reservoir and is the biggest geothermal power plant in China. Yangbajing power plant has been running for more than 30 years, with abundant geological data collected. However, after continuous exploiting for more than 30 years, the heat reservoir has obviously withdrawed, the land subsidence rate is accelerating, and the wellhead temperature, pressure and flowrate have all declined much. Now the shallow reservoir can only support 16 MW capacity, so development of deep high-temperature heat reservoir at this site has very important significance for capacity expanding and sustaining of the ground power plant [7], [10], [11], [12], [13] and [14].

Presently in most conceptual or actual EGSs, the wells are vertical or sub-vertical, and studies on production performance of horizontal well are rarely reported [1]. The horizontal well technologies have advanced rapidly in the past 10 years, and have been used widely in development of subsurface resources, such as shale gas, thickened oil and natural gas hydrate [15] and [16]. Li Gang, Li Bo et al. investigated the production performance of horizontal well in exploiting natural gas hydrates; under the condition that reservoir vertical dimension is small while reservoir horizontal dimension is large, the production performance of horizontal well is obviously better than that of vertical well because the horizontal well can reach and control more reservoir domain [17], [18], [19], [20] and [21]. Zeng et al. reported the heat production performance using two horizontal wells from hot dry rock at Desert Peak site [22].

Research Objectives

This work mainly aims to report the simulation results on electricity generation potential from the 950~1350 m fractured granite reservoir through a single horizontal well at Yangbajing geothermal field. Firstly current development status of deep high-temperature heat reservoir in the north of Yangbajing geothermal filed is reviewed; then based on survey data from well ZK4001, the feasibility to exploiting the 950~1350 m heat reservoir through a single horizontal well is analyzed. A potential heat mining and electricity generating scheme using a single horizontal well is proposed and numerically investigated. We report the research results on electricity generating potential and its main affecting factors over a 20 year period in this study.

In the following sections, we first introduced the detailed geological settings of the Yangbajing geothermal field in Section 2, then we carefully described the methodology for establishing the conceptual and numerical model of electricity generation through a single horizontal well in 3 and 4; in 5 and 6 we reported the results and discussion.

GEOLOGICAL SETTING OF THE YANGBAJING GEOTHERMAL FIELD

The Yangbajing geothermal field is located 94 km northwest of Lhasa, the capital of the Tibet Autonomous Region, China, as shown in Fig. 1. Its geographic coordinate ranges from 30°01'N to 30°05'N and 90°27'E to 90°31'E, with elevation ranging from 4290 m to 4500 m above sea level. It is the first high-temperature hydrothermal convective geothermal field in China [7], [10], [11], [12], [13] and [14]. The mean annual atmospheric pressure is about 0.06 MPa; the mean annual temperature is 2.5 °C with extreme low temperature of −30 °C; the mean measured terrestrial heat flow is about 108 mW/m² [7]. It is inferred that there is a crust molten mass in depth of 5~15 km at the geothermal field and the molten mass is just the heat source for the geothermal field [7]. The geothermal water is mainly from atmospheric precipitation and snow melting, with a general replenishment level of 4860 m [7] and [12].

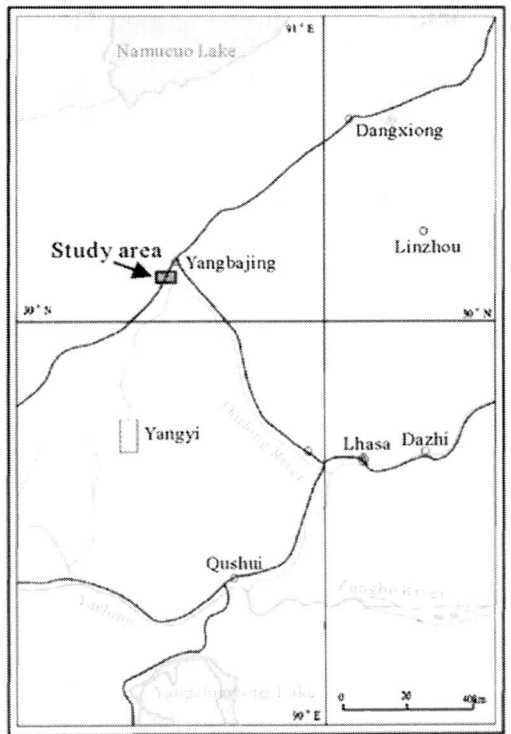

Figure 1: Location of Yangbajing geothermal field in Lhasa, China.

Fig. 2 shows structure and distribution of geological faults and fractures at Yangbajing geothermal field [23]. In active tectonic zones at Qinghai–Tibet Plateau, scope of the Nagqu–Yangbajing–Duoqingcuo active fault belt is very large, and geothermal resource stored here is the most abundant. The Yangbajing geothermal field just locates in a big graben basin (named as Dangxiong–Yangbajing Basin) in the middle of this active fault belt. Southwest of this graben basin is high while northeast is low. This graben basin extends toward northeast as S-shaped, with length about 70 km and width about 7~15 km [10], [11], [12], [13] and [14]. Tectonic activity within Yangbajing geothermal field is very strong, and there widely develop three sets of faults toward northeast, northwest and north-south direction, respectively.

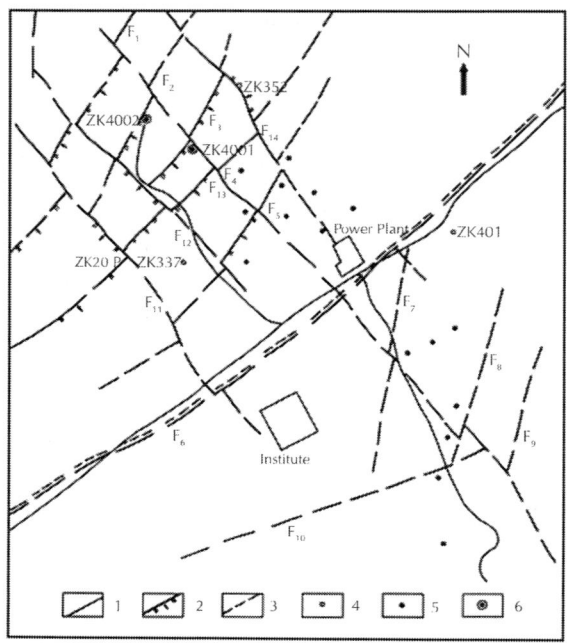

Figure 2: Fault and fracture distribution at Yangbajing geothermal field [7] and [23]. 1-Road; 2-Observerd fault; 3-Buried or inferred fault; 4-Exploration well; 5-Production well; 6-Deep well.

Fig. 3 shows conceptual model of geologic structure of Yangbajing field [23]. Large amounts of thawy water from the Nyainqêntanglha Range and atmospheric precipitation permeate underground along the faults belts, and deep water-bearing strata are continually replenished as heat is transferred from the rock to the water in the cycle. The lower-density hot water produces natural upwelling along the faults, forming a high-temperature thermal storage contained within the relatively closed fissure system, as shown in Fig. 3[7],[12], [13] and [14]; finally there form two types of heat reservoir: shallow Quaternary porous reservoir and fractured granite reservoir in shallow or deep strata. Fluid flow in the shallow reservoir is mainly controlled by the NW faults and fractures [10], [11], [12], [13] and [14]. Both shallow and deep geothermal water is of sodium chloride type [23] and [24].

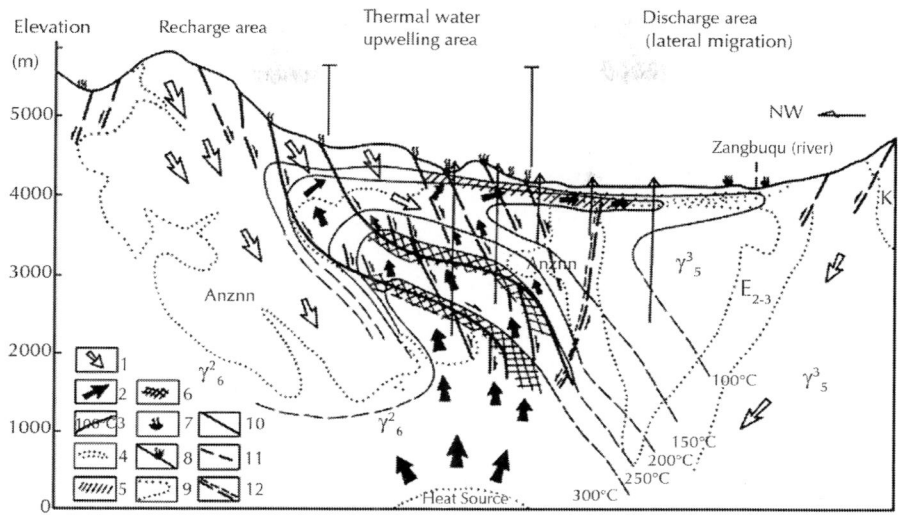

1. Meteoric water. 2. Upwelling thermal water. 3. Isothermal line. 4. Quaternary porous reservoir. 5. Shallow reservoir of fractured granite. 6. Deep reservoir of fractured granite. 7. Hot spring. 8. Steaming ground. 9. Geological boundary. 10. Sliding plane. 11. Normal fault. 12. Buried fault.

Figure 3: Conceptual model of geologic structure of Yangbajing geothermal field, from northwest to southeast [12], [13], [14] and [23].

The Quaternary porous reservoir is the secondary thermal reservoir, and the shallow thermal water is a mixture of the deep thermal water and cold groundwater. The shallow reservoir is located beneath variable thicknesses of politic conglomerate or silty clay in Quaternary conglomerates, glacial sandy-gravel and weathered granite. The basement is Himalayan granite and tuff. The shallow reservoir is generally between depths of 180~280 m corresponding to an elevation of 3800~4020 m above sea level, with the temperature ranging within 150~160 °C and reservoir thickness ranging within 11.8~345.5 m [7] and [23]. Historically, the shallow reservoir is divided by the China–Nepal Highway into Southern and Northern parts. The area of the part of reservoir temperature higher than 40 °C is about 14.62 km², and that of the part of reservoir temperature higher than 130 °C which can be used for electricity generation is about 5.656 km²[7]. The shallow reservoir is the main exploiting zone presently, and the production wellhead conditions are normally 125~140 °C and 0.20~0.46

MPa, with a fluid production rate of 72~169.7 tons/h from a single well [23] and [24].

The deep fractured granite reservoir is expected to have an area of 3.8 km² below a depth of 750 m. It is a typical bedrock-fractured geothermal reservoir. It is proved that the location of the deep reservoir and migration of the deep thermal water are strictly controlled by faults in this region. The geothermal water is stored in fracture zones and tectonic fissures in rocks [25] and [26]. In the northern part, well ZK4001 was drilled to 1450 m and well ZK4002 was drilled to 2006.8 m, both wells are the exploration well for deep heat reservoir [7], [14], [24], [25] and [26]. Fig. 4 shows temperature and pressure logs from well ZK4001 [23]. The first deep fractured granite reservoir is found at depth of 950~1350 m in the well ZK4001 and at depth of 785~1010 m in the well ZK4002 [7], [23] and [24]. This deep reservoir is within mylonitized granite, granitic mylonite and fractured granite that have characteristics of both ductile and brittle shearing, and is covered by intensely altered and impermeable granite, haplophyre and biotite-granite. Feldspars in the granite at shallow depths are replaced by kaolinite to make a more impermeable cap rock. The first fractured granite reservoir has an average temperature of 248 °C [23]. The second deep fractured granite reservoir is found below a depth of 1850 m with temperature greater than 300 °C but no production zone has been penetrated to date from well ZK4002. However, well ZK4002 has no stable flowrate and wellhead conditions during a discharge test but the flowrate of ZK4001 quickly stabilized and the wellhead conditions only varied slightly during a 15-day long discharge test [23], [24] and [25]. It has been demonstrated that ZK4001 and ZK4002 are in the same hydraulic system and the wells are connected to each other [23]. So in this work we mainly research the scheme and feasibility to exploiting the first 950~1350 m fractured granite reservoir through well ZK4001. The 950~1350 m fractured granite reservoir is the upper reservoir of fractured granite as shown in Fig. 3. Due to lack of deep data and simplifying problem, in this work we assume the first fractured reservoir is located all at depth of 950~1350 m, and is conceptually named as the 950~1350 m fractured reservoir. The average temperature of this heat reservoir is 248 °C, and the corresponding pressure is within 8.01~11.57 MPa. This pressure range is computed according to the relationship between pressure P (MPa) and depth z (m) as P = −0.0089z−0.4444 (MPa), which is also indicated in Fig. 4 [23].

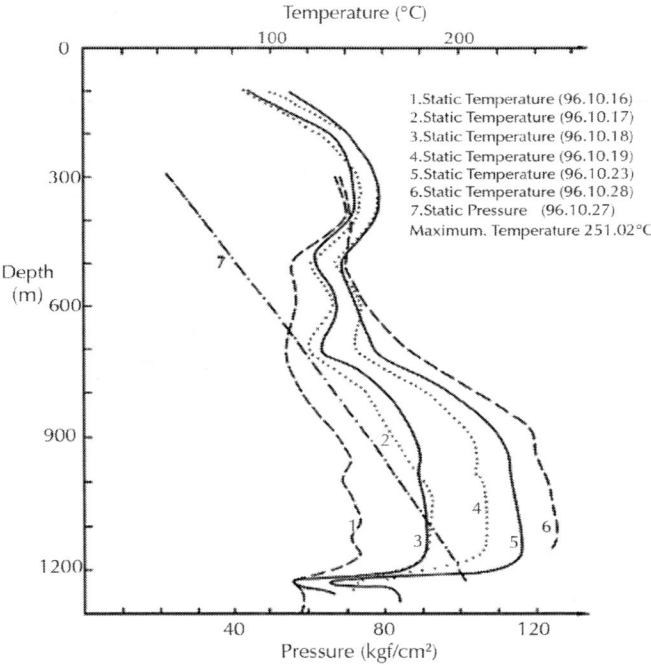

Figure 4: Temperature and pressure logs from well ZK4001 [23] (1 kgf/cm² = 0.1 MPa).

The shallow reservoir at Yangbajing geothermal field has been exploited for more than 30 years, and now in order to expand and sustain the capacity of the ground power plant it rises up the agenda to develop the deep fractured high-temperature heat reservoir [7] and [12]. Zhou et al. proposed to take the natural fault zone near the high-temperature liquating rock region as the location of an artificial reservoir, and through drilling a vertical injection well and two inclined production wells to depth of 8000–9000 m an artificial reservoir of 3×10^{11} m³ in volume can be created at a total cost of about RMB 1 billion Yuan [12], [13] and [14]. Though this scheme reduces the reservoir stimulation cost by using natural fault and fracture zone, the drilling cost is greatly increased because the wells are too deep.

Furthermore, some key geological data such as reservoir lithology, temperature and pressure etc below depth of 8000 m need more exact geological exploration and the risk to establish such a system is very great at this field. As what mentioned before, well ZK4001 and well ZK4002 at north field have shown very great geothermal gradient, and the temperature, thickness, lithology and water chemistry of the 950~1350 m fractured granite reservoir have been clearly investigated. So according to suggestions of Zhao et al. the northern deep reservoir at Yangbajing should be first exploited and then the southern deep reservoir [7].

Based on the existing data, we propose to exploit the 950~1350 m fractured granite reservoir through a single horizontal well near well ZK4001 due to three aspects. First, the well depth of 950~1350 m is much shallower than 8000~9000 m and this will greatly reduce the drilling cost [1] and [16]. Second, the 950~1350 m reservoir is mainly composed of naturally fractured granite and in-situ groundwater, with an intrinsic permeability of about (1~25) mD (1 mD = 1×10^{-15} m²) [25], and no artificial stimulation or only low level stimulation is needed to create the target reservoir [1]. Last, the well ZK4001 is very close to the surface power plant, and this can help to make full use of present surface facilities [7] and [12].

The presence of natural fractures or faults and partial in-situ groundwater makes it much easier to exploit the geothermal energy, such as France Soultz EGS reservoir [1] and [5] and Germany Landau EGS reservoir [27], and the hot wet rock system near a hydrothermal field has become the first choice to exploit the high-temperature rock geothermal energy [1].

SYSTEM DESCRIPTION AND PRODUCTION STRATEGIES

The Fractured Granite Reservoir

As mentioned above, the lithology of the 950~1350 m reservoir rock is varied, that of its impermeable cap rock is also varied, but the main composition of both rock is granite [23] and [25]. For simplification we neglect the variation of lithology and assume that all the cap rock, reservoir rock and base rock are granite [22] and [25]. As shown in Fig. 3, the deep relatively closed fractured reservoirs are located in the thermal water upwelling zone, and continuously receive vertical recharge from deep from macroscopic viewpoint [23] and [25]. The vertical recharge rate over an area of 4158 km² of this geothermal field is about 25 kg/s [25]. To simplify the analysis, for the 950~1350 m reservoir we neglect the vertical recharge within a very limited area from deep during an exploiting period of 20 years, thus the base rock can also be regarded as impermeable. Under such condition both the water and heat production rates are low estimation of practical production rates, but the assumption of impermeable base rock makes the analysis very simple [28] and [29]. In this way the cap rock and base rock are all impermeable. Because there is no water flow within the cap rock and base rock, the pressure within cap and base rock retains unchanged during the heat production period [17] and [29]. Because the horizontal dimension of the reservoir is far greater than the vertical dimension of 400 m, the horizontal well can greatly expand the heat mining zone [17], [18], [19], [20], [21] and [22]. Suppose the 950~1350 m reservoir is located between z = −950 m and z = −1350 m. At depth of 1250 m there is a set of horizontal wells with spacing of 1000 m, which locate at x = −1000 m, 0 m, 1000 m, ..., as shown in Fig. 5. Because of symmetry every well has an exploiting range of 1000 m, and the middle interface of two adjacent wells is no-flow for mass and heat [17], [18], [19], [20] and [21]. In this work we take the well at x = 0 m as the main study object, with a heat exploiting range of 1000 m from x = −500 m to x = 500 m; the boundaries at x = −500 m and x = 500 m are no-flow for mass and heat [29]. This kind of physical model for horizontal wells system has been widely used in previous studies [17], [18], [19], [20] and [21].

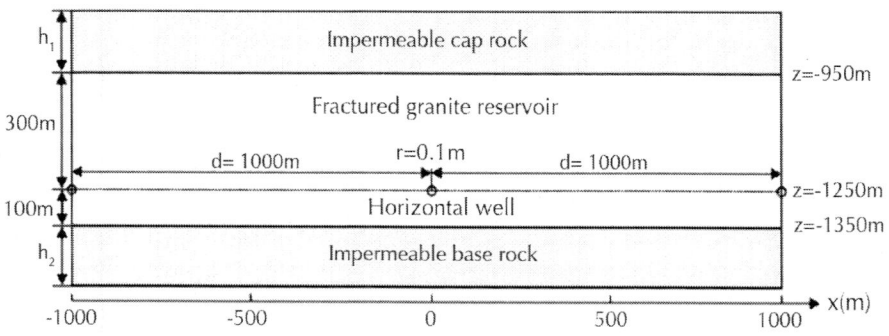

Figure 5: Conceptual model of the 950~1350 m fractured granite reservoir at Yangbajing geothermal field.

Three methods have been used to simulate heat flow in fractured porous rocks: equivalent temperature approach, matrix-fracture temperature approach and rock-fluid temperature approach [30]. Because of simplicity in numerical formulation the equivalent temperature approach has been widely used. Based on the assumption of local thermal equilibrium, the fractured system is represented as a single porous medium and the porous medium temperature is used to represent the temperature at the fracture surface, rock matrix and fluid. However, only when the fracture spacing is less than 2–3 m then the instantaneous local thermal equilibrium between matrix and fracture is a valid assumption [30] and [31]. Pashkevich et al. investigated the performance characteristics of supercritical EGS reservoir with the equivalent temperature method [32], Fakcharoenphol et al. used the equivalent temperature method to study the performance characteristics of EGS reservoir under the coupling condition of water flow, heat transfer and rock deformation

[33]. Radilla et al. presented a new interpretation of tracer tests in the EGS of Soultz using the equivalent stratified medium approach [34]. Birdsell et al. developed a three-dimensional model of the Fenton Hill EGS reservoir using equivalent porous medium, which matched hydraulic, thermal, tracer and water loss data during a 30-day flow test [35]. Pruess et al. compared the simulation results among a semi-analytical approximation method, the MINC (multiple interacting continua) method and the porous medium method [36]. McDermott, Watanabe et al. detailedly reported the effect of coupling interaction of fluid flow, heat transfer and rock deformation on heat production performance of EGS reservoir [37] and [38]. Sanyal et al. pointed out that after hydrofracturing the fracture spacing normally ranges within 0.33~300 m [28], the average fracture spacing of Fenton Hill EGS and Spa Urach EGS are all proved to be about 1 m [35] and [37], and that of Rosemanowes EGS is proved to be several meters [39]. Based on these facts in this work we assume that after stimulation through the horizontal well the fracture spacing is less than 2 m [22], [28], [35] and [37], this fracture distribution condition has been proved real in Fenton Hill, Spa Urach and Rosemanowes EGS fields [35], [37] and [39], and this fracture condition can also strictly meet the application requirement of equivalent porous medium method [22] and [31]. In this way the 950~1350 m fractured granite reservoir can be represented as a single porous media [22], [30], [31], [35], [36], [37], [38] and [39]. The coupling interaction of fluid mechanical, rock mechanical, hydraulic, thermal and chemical processes will affect the heat production performance of EGS, however, Mcdermott et al. proved that under different conditions regarding different coupling processes, the heat production rate is minimum when water properties are functions of temperature, pressure and salinity while rock properties are constant, without considering the mechanical interaction between rock and fluid as well as the rock thermoelastic effect [22] and [37]. Once the fluid–rock interaction or rock deformation is considered, the heat production rate will be increased [37]. For simplification in this work we have assumed that after stimulation the fracture characteristics, such as fracture spacing and aperture maintain unchanged over the heat

production period of 20 years [22], [28], [30] and [36]. In this way the simulated heat production rate is the lower limit of the system thermal output. Due to lack of practical geological data, in this work we just assume that this reservoir has been created for heat production, and the detailed stimulation method for creation such a reservoir is not discussed [22].

The Single Horizontal Well

Li G and Li B et al. discussed the performance characteristics of a single horizontal well in exploiting natural gas in detail [17], [18], [19], [20] and [21]. Zeng et al. reported the performance of two horizontal wells in exploiting the HDR geothermal energy at Desert Peak geothermal field [22]. According to the experience from oil and gas industry, the adopted horizontal well is shown in Fig. 6. The horizontal well is located at x = 0 m and z = −1250 m, and its radius r is r = 0.1 m, as shown in Fig. 5. There are 8 grooves evenly distributed along the circumference of the horizontal well. The produced thermal water flows into the well through the grooves and then is extracted out and pumped into the power plant by the ground production pump [7] and [22]. This kind of well design has been proved simple and technologically feasible in oil and gas industry [17], [18], [19], [20], [21] and [22]. Because of symmetry only the left or the right side of the well, namely 4 grooves need being simulated. According to the experience from oil and gas industry, the heat production problem using horizontal wells can be reduced to two dimensional in vertical plane [17], [18],[19], [20], [21] and [22]. Suppose the direction along the horizontal well as y. Therefore, in this work only 10 m thickness subdomain of the reservoir along y direction is taken as the simulated domain [22]. If length of the horizontal well is designed as 1000 m, the whole reservoir domain controlled by this single horizontal well will be 1000 m × 400 m × 1000 m; the actual system water and heat production rate is 100 times that of the simulated domain.

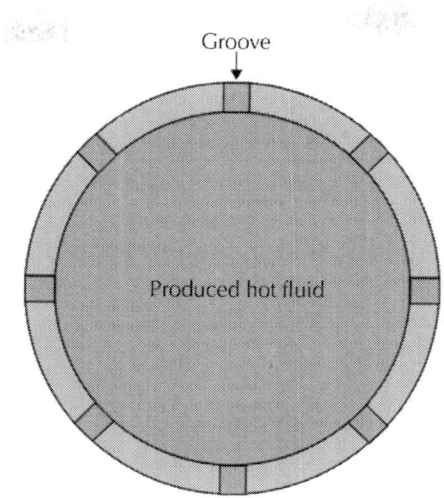

Groove

Produced hot fluid

Figure 6: Well design used for the heat production at the Yangbajing geo-thermal field.

The Heat Production and Electricity Generation Method

In this work, we research the constant flowrate method to exploit the heat energy from the 950~1350 m reservoir through a single horizontal well near well ZK4001. During the heat mining period of the 20 years, the water production rate is maintained at a constant q. The production pump is installed on the ground to extract the in-situ geothermal water out of the fractured reservoir and pump it into the double-flash power plant at Yangbajing geothermal field [7].

When the reservoir geothermal water is mined out by the production pump, the bottomhole pressure P_w along with the reservoir pressure P will decline gradually. Once P_w decreases to lower than the corresponding saturated vapor pressure, the liquid water will begin to vaporize. The initial bottomhole pressure P_{w0} at depth of 1250 m is P_{w0} = 10.68 MPa and the corresponding saturated vapor pressure to the initial average reservoir temperature of 248 °C is 3.84 MPa [40]. Because the initial reservoir pressure is within 8.01~11.57 MPa, all the initial geothermal water is liquid. If the water production rate is too great, the reservoir pressure will decline very quickly, and this will easily arouse vaporization and two phase liquid-steam flow in the reservoir and wellbore. Based on the operational experience from Yangbajing geothermal power plant, the two phase flow will not only greatly increase the instability of reservoir–wellbore flow and pressure loss of fluid transportation in the wellbore and duct, but also arouse precipitation of $CaCO_3$ and the induced blocking of reservoir and wellbore [29] and [41]. The vaporization in the reservoir or wellbore will greatly decrease the production efficiency, even stop the heat production process. Consequently it's better to control the wellbore flow as liquid and maintain the water production rate at reasonable level. In reference case in this work, the water production rate q is 20 kg/s. currently the maximum water production rate of pump has reached 126 L/s, and the production rate of 20 kg/s is technologically feasible [28]. Because the horizontal well length is 1000 m, the water production rate of 10 m length well is 0.20 kg/s.

When the production pump on the ground extracts the reservoir water (liquid or steam) at a rate of q from the depth of h, the electric power of the pump W_p is as follows:

$$W_p = \frac{q(\rho g h - P_{pro})}{\rho \eta_p}$$

$$(1)$$

where η_p = 80% is the pump efficiency [22] and [43], P_{pro} >0 the bottomhole pressure, ρ the fluid density (kg/m³), and g = 9.80 m/s² the gravity. In this work h = 1250 m. In long-term fluid production, the heat transfer between the wellbore fluid and surroundings can be neglected, and the wellbore flow can be regarded as isenthalpic [22] and [44]. The bottomhole produced specific enthalpy h_{pro} can be computed as $h_{pro} = h(T_{pro}, P_{pro})$, where T_{pro} is the production temperature. The Yangbajing geothermal power plant comprises single-flash power plant and double-flash power plant, which are all open loop system [41]. The exhaust steam from turbine is condensed to liquid. The condensate water is from the Duilong River (see Fig. 1), with an average temperature of 8~9 °C. For simplification, in this work we assume that the heat rejection temperature of Yangbajing power plant T_0 is T_0 = 9 °C = 282.15 K. After condensing the tail water is directly discharged to Duilong River with a temperature of 66~80 °C [45]. For simplification in this work the average temperature of the tail water is assumed as 70 °C; the local annual mean atmospheric pressure is about 60.59 kPa [41], so the mean residual specific enthalpy h_0 ish$_0$ = 293.04 kJ/kg. During the liquid flash process, partial water may be lost. As stated in Section 3.1, the water production rate in this work is the minimum because the vertical recharge from deep is neglected. So here we neglect the slight water loss in the flash process and assume all the liquid water is flashed to steam. In this way based on the law of thermodynamics, the thermal water at flowrate of q can generate electricity at electric power of W as follows [22] and [28]:

$$W = 0.45q(h_{pro} - h_0)\left(1 - \frac{T_0}{T_{pro}}\right)$$

(2)

where $(1-T_0/T_{pro})$ is the maximum efficiency of heat converted to available work and 0.45 is the utilization efficiency of available work converted to electric power [22] and [28]. Consequently, the energy efficiency η of heat mining and electricity generation through a single horizontal well is:

$$\eta = \frac{W}{W_p} = \frac{0.45\eta_p(h_{pro} - h_0)\left(1 - T_0/T_{pro}\right)}{gh - P_{pro}/\rho}$$

(3)

It can be readily found that the energy efficiency mainly depends on the production temperature T_{pro}, the bottomhole pressure P_{pro} and the fluid density ρ. Higher T_{pro} or higher P_{pro} or lower ρ will result in higher energy efficiency. It is worth noting that the energy efficiency has nothing with the water production rate q, and this is similar to that in the two horizontal wells system [22]. Because there is no cold water injected into the reservoir to mine the heat in the rock and only thermal water is extracted out of the reservoir, in this work we define the recovery factor of geothermal water as fraction of in-place hot water recovered [1] and [28]. Under the condition of only mining the in-place hot water, the recovery factor r of hot water during a period of t is as follows:

$$r = \frac{qt}{\rho V \phi}$$

(4)

where ϕ is reservoir porosity, V total reservoir volume, t total mining time. Obviously, if there is no vaporization, a higher q will result in a higher recovery factor of r within a constant mining period, along with a greater electric power W; there will be more in-situ hot water produced out.

NUMERICAL MODELS AND SIMULATION APPROACH

Domain, Grid and Parameters

In this work, we used the TOUGH2-EOS1 code to simulate the heat production process during a 20 year period [22] and [46]. This code can calculate the bottomhole (T_{pro}, P_{pro}) versus time and h_{pro} versus time according to $h_{pro} = h(T_{pro}, P_{pro})$. Based on these and equations (1), (2) and (3), W_p, W and versus time are all calculated. The EOS1 module computes the water density, viscosity, specific enthalpy and saturated vapor pressure based on their functions of temperature and pressure, while for solid rock the properties are constant. As stated above under this condition the heat production, along with thermal power and electric power is minimum [22] and [46].

The study domain locates within $-500 \text{ m} \leq x \leq 500 \text{ m}$ in Fig. 5. Because of symmetry only the $0 \text{ m} \leq x \leq 500 \text{ m}$ part needs simulation. In order to accurately simulate the temperature and pressure field, the thickness of the cap rock and base rock must be great enough to avoid the boundary effect [22]. There is only conductive heat exchange between the impermeable cap rock, base rock and the permeable reservoir. The rock heat diffusivity is very small, about in the order of 10^{-6} m^2/s; furthermore, as the fracture spacing is less than 2 m, the heat transfer area within the reservoir is much larger than the reservoir boundary area. So this conductive heat exchange from cap rock and base rock can be neglected [42] and [47], and top and bottom boundaries of the reservoir are no-flow for mass and heat. In this way, the simulated domain locates within -1350 m $\leq z \leq -950$ m in vertical direction and 0 m $\leq x \leq$ 500 m in xdirection, with all boundaries no-flow for mass and heat. In this work, only 10 m length of the horizontal well within $0 \leq y \leq 10$ m is simulated, and the heat and electricity production of whole length of 1000 m are 100 times that of the simulated section.

Fig. 7 displays grid system for above simulated domain, in which the horizontal well locates at x = 0 m and z = -1250 m. Because heat transfer and water flow around the well changes rapidly, the subdomain within 0 m $\leq x \leq$ 100 m and -1300 m $\leq z \leq -1200$ m (around the horizontal well) is discretized with refined grid blocks of 2 m $\times$ 2 m; the other subdomain, such as within 100 m $\leq x \leq$ 500 m, -1350 m $\leq z \leq -1300$ m and -1200 m $\leq z \leq -950$ m is discretized with grid blocks of 2 m $\times$ 5 m or 5 m $\times$ 5 m. The grid blocks of 2 m $\times$ 2 m, 2 m $\times$ 5 m and 5 m $\times$ 5 m are fine enough to simulate the mass and heat flow process in geothermal system compared with previous studies [22], [42] and [46]. In this way in y direction there is only one cell, in x direction there are 130 cells and in z direction there are 110 cells. The 2-D grid in Fig. 7comprises total 14,300 elements. The flowrate through the horizontal well is evenly assigned into the well elements in which the well locates [46]. Because half of the well locates in two elements of 2 m $\times$ 2 m, one well element will represent the flowrate through two grooves in Fig. 6, namely represent one fourth flowrate of the well. In the reference case, the water production rate of the 10 m length well is 0.20 kg/s, so production rate of every well element is 0.05 kg/s [22].

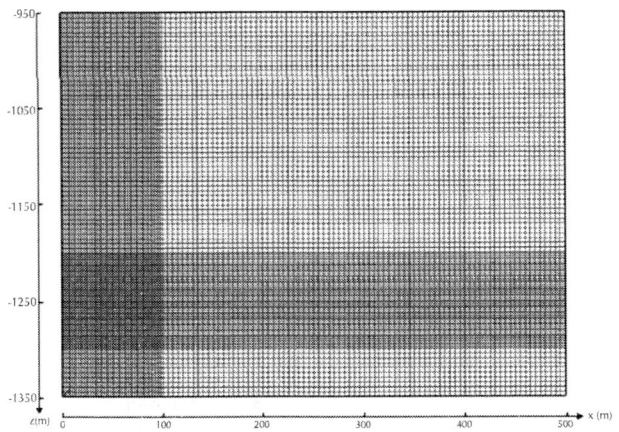

Figure 7: Grid used for domain discretization and numerical computation.

Table 1 shows the properties and conditions in this simulation work [25]. The permeability of the fractured reservoir greatly affects the heat production performance, and Fan confirmed that the intrinsic permeability of deep reservoir at Yangbajing field is within (1~25) mD [25]. The fractured reservoir permeability is closely related to the apertures of the fractures and average spacing between fractures [39]. Sanyal pointed out the fractured reservoir permeability may be within (1~100) mD [28]. The Fenton Hill EGS test proved that average permeability of 1 mD or less will result in great reservoir impedance, very unfavorable for heat production [28] and [35]. Zeng considered the reservoir permeability within (10~100) mD as suitable for effective heat production [22]. In the reference case, the average permeability is assumed as $k_x = k_y = k_z = 10$ mD. Due to lack of actual deep geological data, the detailed reservoir creation process and method is not discussed in this work. For simulation the vaporization process aroused by the pressure decline, we adopt Corey's Curves to compute the relative permeability of liquid phase and vapor phase [46] and [47]. That is, the relative permeability of liquid phase krl is (5) and that of gas (steam) phase krg is (6):

$$k_{rl} = s^4 \tag{5}$$

$$k_{rg} = (1 - s)^2 \left(1 - s^2\right) \tag{6}$$

where s is a parameter, which is computed with $s = (s_l - s_{lr})/(1 - s_{lr} - s_{gr})$; s_l is the liquid phase saturation; s_{lr} and s_{gr} are two model parameters. In this work, $s_{lr} = 0.30$, $s_{gr} = 0.05$ [46] and [47].

Table 1: 950~1350 m reservoir properties and conditions at well ZK4001 at Yangbajing field [25]

Parameter	Value
Rock thermal conductivity	2.50 W/(m·K)
Rock specific heat	1000 J/(kg·K)
Rock density	2650 kg/m³
Reservoir height	400 m
Reservoir length	1000 m
Total well length	1000 m
Simulated well length	10 m
Reservoir porosity (reference case)	10%
Horizontal intrinsic permeability kx = ky (reference case)	10×10^{-15} m²
Vertical intrinsic permeability kz (reference case)	10×10^{-15} m²
Water production rate (reference case)	20 kg/s
Partial water production rate (simulated section)	0.2 kg/s
Initial temperature	248 °C
Initial pressure	P = −0.0089z−0.4444(MPa)
Initial liquid saturation	1

Boundary and Initial Conditions

As stated above, in this work the slight conductive heat transfer between the impermeable cap rock or base rock and the reservoir is neglected [22], [42] and [47]. Because of symmetry and impermeable overburden or underburden, all the boundaries of the simulated domain within 0 m ≤ x ≤ 500 m and −1350 m ≤ z ≤ −950 m are no-flow for mass and heat. As shown in Table 1, the initial reservoir temperature is 248 °C; the initial reservoir pressure is P = −0.0089z−0.4444 (MPa). The corresponding saturated vapor pressure to liquid water at 248 °C is 3.84 MPa, so initial liquid saturation is 1.

ELECTRICITY GENERATION THROUGH A SINGLE HORIZONTAL WELL

The Reference Case

As shown in Table 1, in the reference case the intrinsic permeability of the reservoir is $k_x = k_y = k_z = 10$ mD, the fluid production rate of 10 m length well is 0.2 kg/s and total fluid production rate is 20 kg/s. In this work we performed the heat production and electricity generation for a period of 20 pears.

Production Temperature, Specific Enthalpy and Steam Saturation

Fig. 8 shows the evolution of production temperature T_{pro} and specific enthalpy h_{pro} during 20 years. We can see that based on the change of T_{pro}, the production process can be divided into two stages: stable stage and declining stage. During the stage of 0~19.20 years, variation of T_{pro} is very slight, T_{pro} basically retains at 247.8 °C, slightly less than initial temperature of 248.0 °C and this is the stable stage. In this stage, h_{pro} basically retains at 1075.20 kJ/kg, also with very slight variation. Fig. 9 shows the evolution of steam saturation in the produced fluid. It can be readily found in the stable stage the system only produces liquid water, with production steam saturation at 0. During the stage of 19.20~20 years, T_{pro} declines from 247.8 °C to 243.8 °C, while h_{pro} increases from 1075.20 kJ/kg to 1145.18 kJ/kg. This is because during the declining stage, P_{pro} has decreased to lower than the saturated vapor pressure at 248.0 °C of 3.84 MPa as shown in Fig. 10, the bottomhole liquid water begins to vaporize; the production steam saturation increases from 0 to 0.28, shown as in Fig. 9, and the system produces the mixture of liquid and steam. During the declining stage, both P_{pro} and T_{pro} are decreasing with time, and the increasing of h_{pro} with time is aroused by latent heat of vaporization of water; the more the steam saturation in the production fluid, the higher the latent heat of vaporization contained in the production fluid. Furthermore, the latent heat of vaporization of liquid water increases with decreasing pressure. So, on the whole the increasing of h_{pro} during declining stage is caused by the decreasing of P_{pro}. Correspondingly, the vaporization is an endothermic process, and this causes the declining of T_{pro} during the declining stage. As stated above, the two-phase liquid steam flow in the wellbore will result in great pressure loss for fluid transportation, and arouse chemical deposition in the reservoir and high flow impedance [41]. In the reference case, the declining stage only lasts for about 0.80 year, and the two-phase flow effect at Yangbajing geothermal field may be very slight. Compared with the two horizontal wells system, main difference is that P_{pro} will decrease all through the production process, and this will cause the increasing of steam production and h_{pro} in the declining stage [22]. Meanwhile, because there is no cold water injected into the reservoir, most heat energy stored in the rock is not extracted out, and T_{pro} only declines very small. So, the heat recovery factor is much lower than that of two wells system [22]. (The detailed parameters of the two wells system have been listed in Refs.[22], and that of the single well system have been listed in Table 1 in this work.)

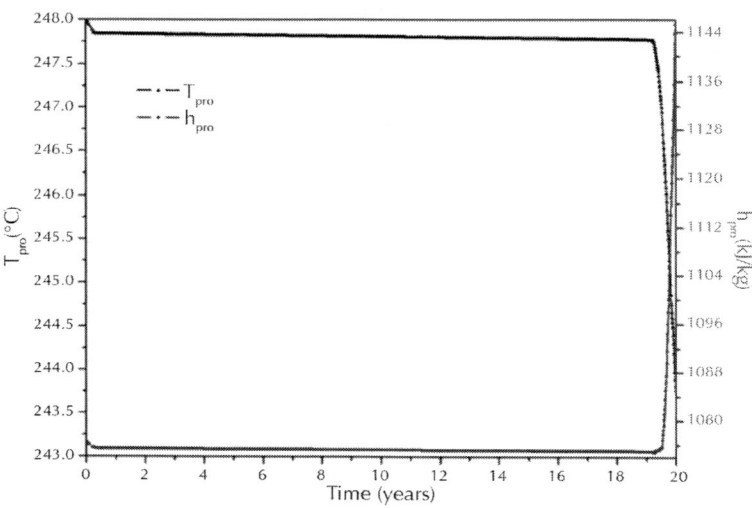

Figure 8: Evolution of production temperature and specific enthalpy during the 20 years.

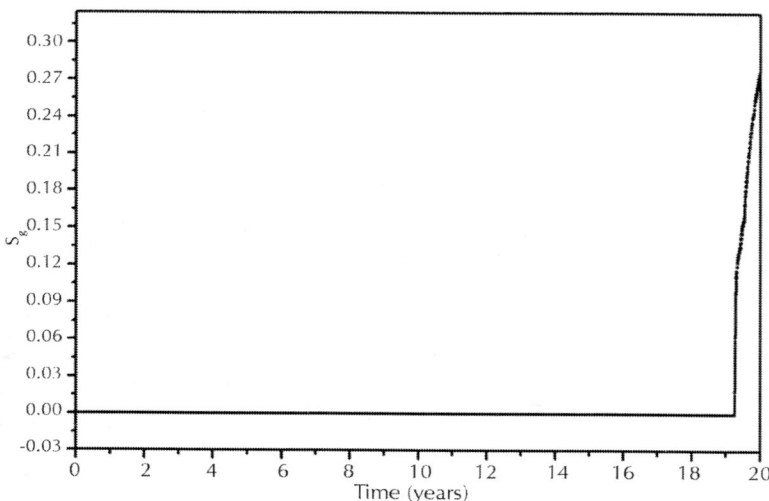

Figure 9: Evolution of production steam saturation during the 20 years.

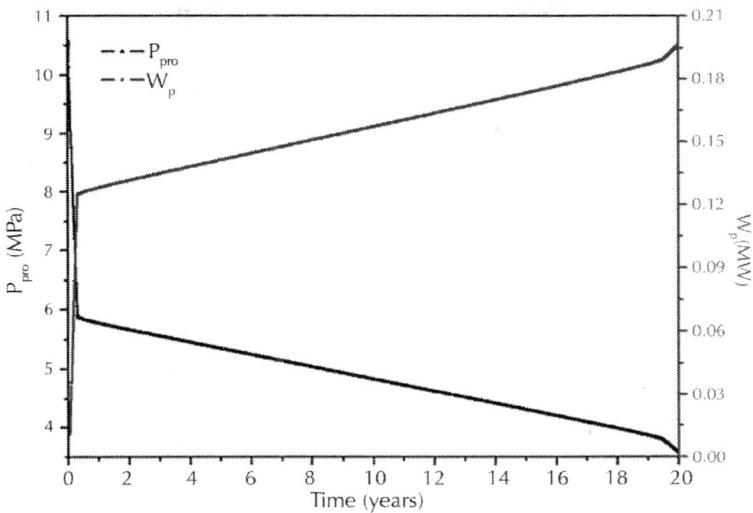

Figure 10: Evolution of bottomhole pressure and pump power during the 20 years.

Bottomhole Pressure and Pump Power

Fig. 10 shows the evolution of bottomhole pressure and pump power during 20 years. In all the calculations with equations (1), (2), (3) and (4), we adopt the maximum water density of = 810 kg/m³, which will result in a maximum of W_p and a minimum of , making the analysis more valuable. We can find at the beginning of the production there is a sharp drop of P_{pro} in Fig. 10 from the initial pressure to a relative stable pressure of 8.76 MPa. Because this period lasts very short and P_{pro} changes too quickly, there may be some uncertain calculation errors in the simulation. To be more accurate we neglect this very short period of fast declining of P_{pro}; the same applies to the analysis of W_p and . During the stable stage of 0~19.20 years, P_{pro} gradually decreases from 8.76 MPa to 3.84 MPa, and the corresponding W_p increases from 0.04 MW to 0.19 MW. As shown in Fig. 9, the system only produces liquid water during this stage. During the following stage of 19.20~20 years,

P_{pro} decreases from 3.84 MPa to 3.58 MPa, in which range the liquid water begins to vaporize in the wellbore; the corresponding W_p gradually increases from 0.19 MW to 0.20 MW, and the system produces the mixture of liquid and steam. The change characteristic of P_{pro} during the production process is similar to that of two vertical wells system at Desert Peak geothermal filed, and this can support the simulation results to some extent [48].

Electric Power and Energy Efficiency

Fig. 11 shows the evolution of electric power W and energy efficiency η during the 20 years. During the stable stage, W basically retains at 3.23 MW; η decreases from about 50.00 to 17.16 as a result of continuous increasing of W_p. During the declining stage, because h_{pro} is slowly increasing, W also gradually increases from 3.23 MW to 3.48 MW; W_p is within 0.19~0.20 MW, and the energy efficiency η gradually increases from 17.16 to 17.78. When comparing these with the two horizontal wells system, both electric powers maintain unchanged and energy efficiency gradually declines during the stable stage; during the declining stage, both the electric power and energy efficiency continuously declines for the two horizontal wells system, while due to production of steam the electric power along with the energy efficiency is slightly increased for the single horizontal well system [22]. Evans et al. proposed that a two-well EGS should produce 3.50 MW of electric power at a fluid production rate of 50 kg/s [4], while in this work the single well system produces an electric power of 3.23~3.48 MW at a fluid production rate of 20 kg/s. If we consider the electric power generated by unit fluid production rate, the efficiency of this single well system well meets the industry requirement. Zeng et al. found an ideal energy efficiency of the two horizontal wells system is within 30.6~10.8 [22], in this work the energy efficiency is within 50.00~17.16, so basically the energy efficiency of the single well system also meets the industry requirement. Because there is enough in-place reservoir fluid, there is no need to drill injection well and circulate the fluid throughout the fractured reservoir, and the construction and operation of the single well system is much easier [1] and [22].

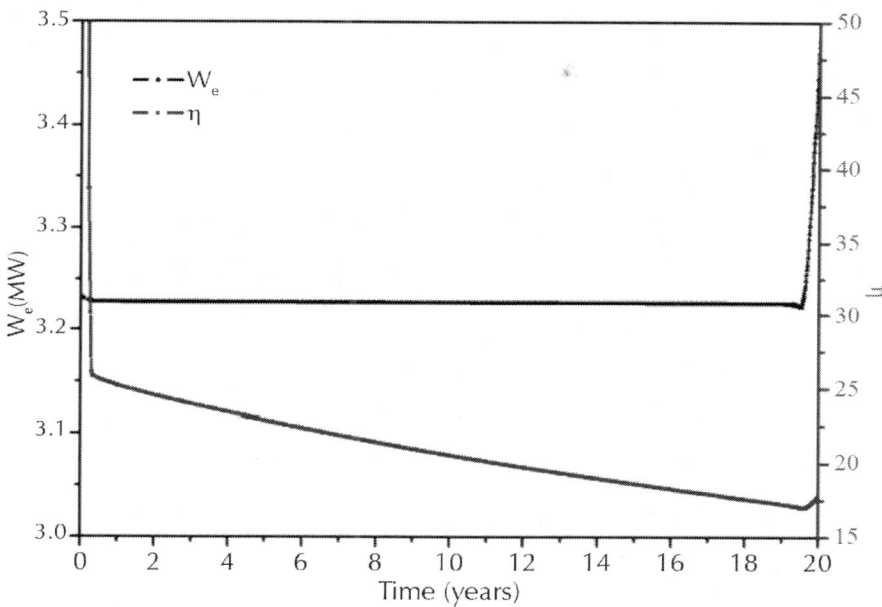

Figure 11: Evolution of electric power and energy efficiency during the 20 years.

Recovery Factor of Geothermal Water

In the reference case, the recovery factor of geothermal water in this work during 20 years is about 38.93% according to equation (4), in which the fluid density has been assigned the maximum value of 810 kg/m³, with reservoir volume of $V = 4 \times 10^8$ m³ and $= 10\%$. It is worth noting that the produced heat is mainly from the in-place geothermal water, in the rock there is only little heat extracted out under the condition of no cold water injection, so the in-place heat recovery factor is certainly less than 38.93% in this study [22].

Spatial Distribution of Temperature

Fig. 12, Fig. 13 and Fig. 14 show the evolution of spatial distribution of temperature, pressure and steam saturation during the 20 years, respectively.

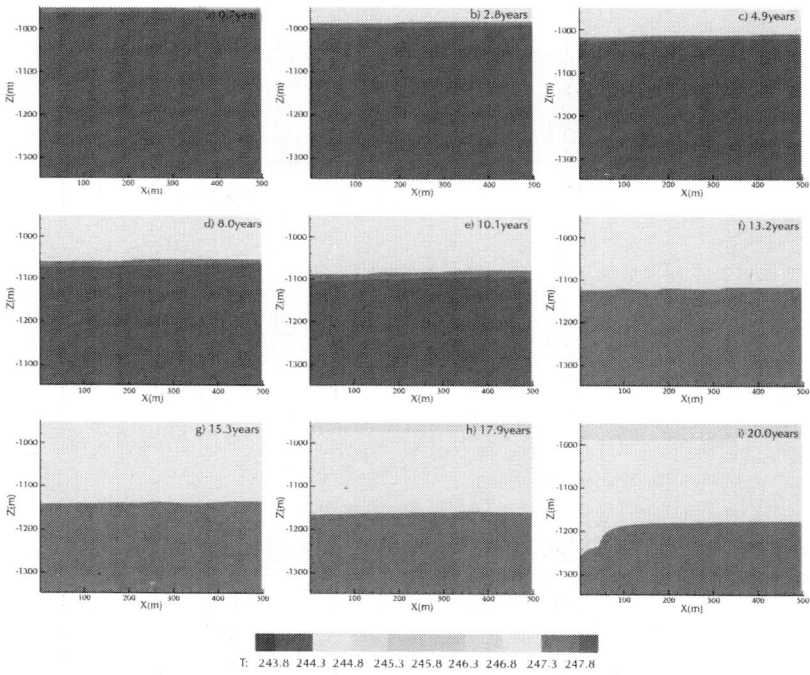

T: 243.8 244.3 244.8 245.3 245.8 246.3 246.8 247.3 247.8

Figure 12: Evolution of the spatial distribution of temperature (°C) during the 20 years.

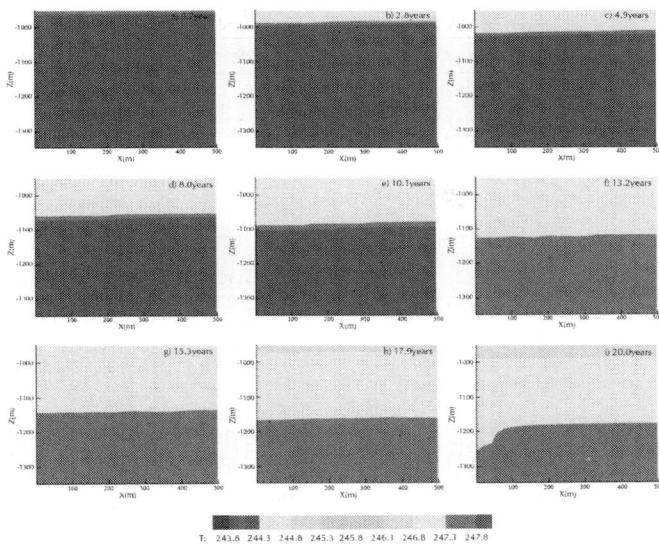

Figure 13: Evolution of the spatial distribution of pressure (MPa) during the 20 years.

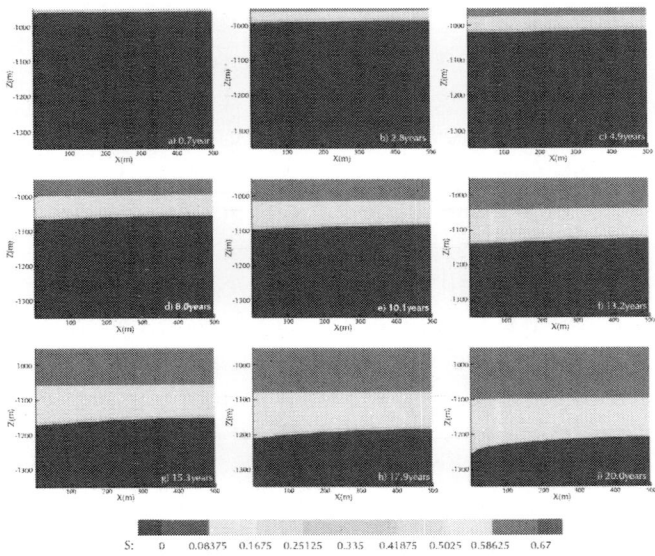

Figure 14: Evolution of the spatial distribution of steam saturation during the 20 years.

Fig. 12 shows the evolution of spatial distribution of temperature during the 20 years. As stated above, the slight conductive heat transfer from the impermeable cap rock and base rock is neglected in this work [22], [42] and [47], and the top and bottom boundaries of the reservoir are no-flow for mass and heat. The figure approximately displays the shape and movement of interface between different temperature zones. The temperature interfaces are all slightly inclined, migrate downward as heat production, so they provide a representation and measure for the heat production process. Since the heat production begins, the reservoir temperature changes to be uneven quickly, and the temperature slightly increases with increasing depth. As the heat production goes on, the temperature interfaces gradually migrate downward and the upper low temperature zone continuously expands (Fig. 12a–h). Before the characteristic temperature interface arrives at z = −1250 m where the horizontal well locates, the production temperature T_{pro} always retains at 247.8 °C, and this period corresponds to the stable stage; once the characteristic temperature interface arrives at the horizontal well, the fluid from the upper lower temperature zone is strongly extracted out by the well, T_{pro} begins to decline and this period corresponds to the declining stage (Fig. 12i). Comparing Fig. 12 with Fig. 14, we can find the upper low temperature zone is close to the upper steam contained zone, and the upper low temperature zone is formed by the vaporization which absorbs heat. The vaporization induced by pressure drop is an endothermic process, and will cause a local redistribution of the heat energy. However, all over the reservoir the temperature decline is very small, only several Celsius degrees. It can be inferred that compared with the cold water injection system, the only in-place hot water production system mainly produce the heat energy in the thermal water, leading to a very low heat recovery factor during the period of the 20 years [22] and [28].

Spatial Distribution of Pressure

Fig. 13 shows the evolution of spatial distribution of pressure during the 20 years. The saturated vapor pressure of liquid water at 248 °C is about 3.84 MPa. When reservoir pressure declines to lower than 3.84 MPa due to hot water mining, partial liquid water will vaporize and form two-phase steam liquid zone. We can find the pressure interfaces between different zones are also inclined, and migrate downward as heat production, so they also provide a representation of heat production. The reservoir can be divided into two zones if we take 3.84 MPa as a pressure boundary. The upper zone is low pressure which basically retains at 3.84 MPa, the liquid water continuously vaporize to steam, so this is a two-phase steam liquid flow zone, as shown in Fig. 14. The lower zone is high pressure than 3.84 MPa, in which there is only liquid water, and the pressure increases with increasing depth. As the heat production goes on, the pressure interfaces continuously migrate downward, and in the upper zone the pressure basically retains unchanged while in the lower zone the pressure gradually declines (Fig. 13a–h). After the characteristic interface arrives at $z = -1250$ m, P_{pro} declines to lower than 3.84 MPa, the well begins to produce the mixture of liquid and steam, and this period corresponds to the declining stage (Fig. 13i). These characteristics are according with those as shown in Fig. 10.

Spatial Distribution of Steam Saturation

Fig. 14 shows the evolution of spatial distribution of steam saturation during the 20 years. Similar to Fig. 12 and Fig. 13, the steam saturation interfaces between different zones are slightly inclined, and the interfaces migrate downward as the heat production. This provides a measure of the heat production process. The reservoir can be divided into two zones if we take 0 as the steam saturation boundary. In the upper zone, the steam saturation increases with decreasing depth; with more liquid in the lower zone vaporize to steam, the upper two-phase steam liquid zone gradually expands with time. In the lower zone, the pressure gradually declines, and more and more liquid water in the boundary between two phase zone and liquid phase zone vaporize to steam, making the liquid zone gradually diminished. Before the characteristic interface arrives at z = −1250 m where the well locates, the well is always located in the liquid zone, the system only produce liquid water, and this period corresponds to the stable stage (Fig. 14a–h); after the characteristic interface arrives at the horizontal well, the upper zone is strongly extracted out by the well, the well produces the mixture of liquid and steam, and this period corresponds to the declining stage (Fig. 14i).

It is worth noting that the steam saturation in the upper zone is not evenly distributed, the steam saturation decreases with increasing depth. This is because the pressure increases with the increasing depth, and this increases the difficulty of vaporization with the increasing depth. In the two wells system, the bottomhole production pressure can be easily controlled at higher than saturated vapor pressure which makes all the reservoir fluid liquid, and this makes the operation of the system more stable and much easier [22]. However, for the single well system, we must conduct some investigations to find reasonable fluid production rate in order to avoid immature vaporization in the wellbore and reservoir [41].

Spatial Distribution of Liquid Density

Based on $\rho = \rho(P,T)$, the spatial distribution of liquid density can be calculated. From above we know that during the 20 years the reservoir temperature is within 243.8~248.0 °C and the reservoir pressure is within 3.58~11.57 MPa, so the liquid density is within 810~803 kg/m^3 based on the property tables of water and steam in this work [40].

The variation of liquid density is very small, this is because when pressure ranges within 3.58~11.57 MPa, the liquid water density is much more obviously influenced by the temperature [40]; the temperature only decreases by several Celsius degrees, so the variation of liquid density is also quite small. In all the calculations in equations (1), (3) and (4) above, we adopted the maximum of ρ = 810 kg/m^3 and this results in a maximum pump power and minimum energy efficiency and geothermal water recovery factor. This assignment of liquid density in (1), (3) and (4) makes the analysis more reliable.

SENSITIVITY ANALYSIS OF ELECTRICITY GENERATION TO VARIOUS PARAMETERS

In this work, we have mainly researched the sensitivity of electricity generation through a single horizontal well to the following parameters: the reservoir porosity φ, the rock heat conductivity λ, the horizontal intrinsic permeability $k_x = k_y$, the vertical intrinsic permeability k_z and the fluid production rate q. Based on the reference case (r), the following 5 scenarios are studied: (a) increasing φ to φ = 20%; (b) increasing λ to λ = 3.0 W/(m·K); (c) increasing $k_x = k_y$ to $k_x = k_y$ = 20 mD; (d) increasing k_z to k_z = 20 mD; (e) decreasing q to q = 15 kg/s. Strictly speaking, maybe the sensitivity of electricity generation to the system parameters is nonlinear, and this makes the sensitivity analysis prolix and complicated, which is outside the scope of this work to some extent. According to the experience from similar studies, in this work we don't consider the nonlinear effect of the sensitivity analysis and just compare the reference case with the scenario which changes only one parameter [18] and [22]. Fig. 15, Fig. 16, Fig. 17, Fig. 18, Fig. 19 and Fig. 20 show the sensitivity of production specific enthalpy h_{pro}, steam saturation S_g, bottomhole pressure P_{pro}, pump power W_p, electric power W and energy efficiency η to various parameters, respectively. As stated above, in all the sensitivity analysis, the liquid density in equations (1), (3) and (4) is assigned as ρ = 810 kg/m^3 and this makes the analysis more reliable.

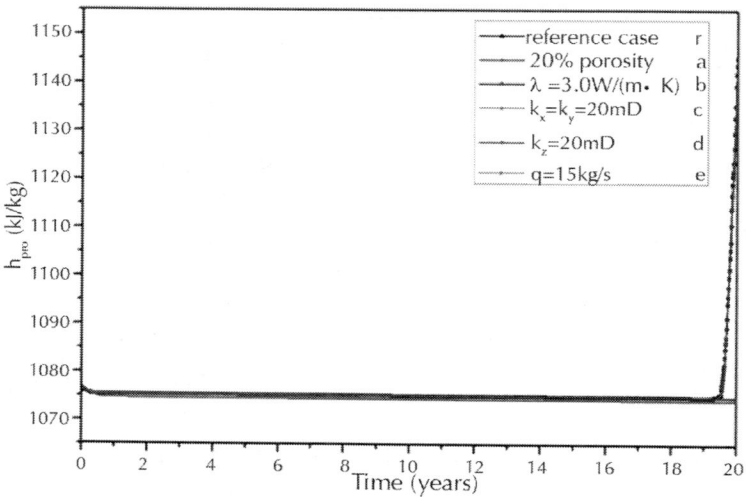

Figure 15: Sensitivity of production specific enthalpy to various parameters.

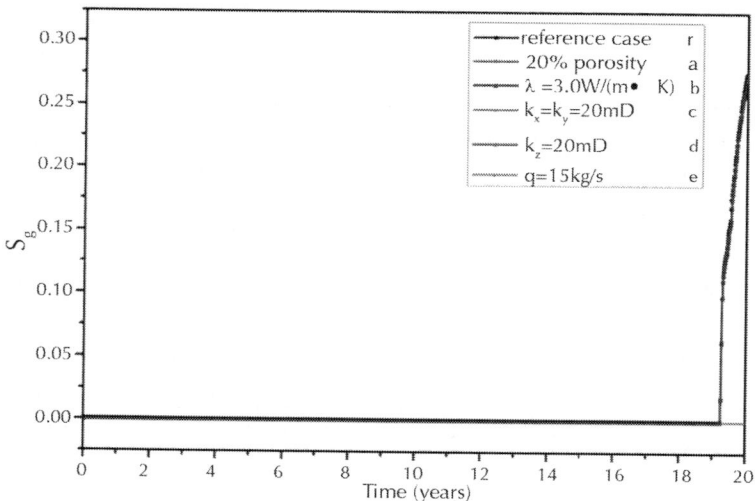

Figure 16: Sensitivity of production steam saturation to various parameters.

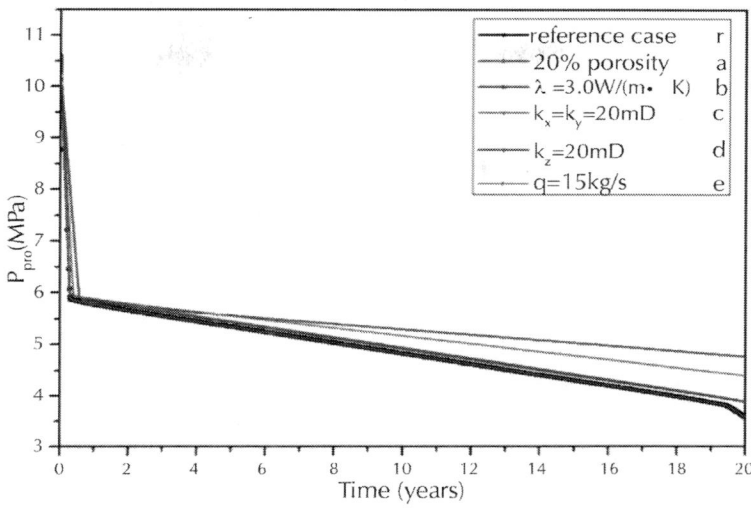

Figure 17: Sensitivity of bottomhole pressure to various parameters.

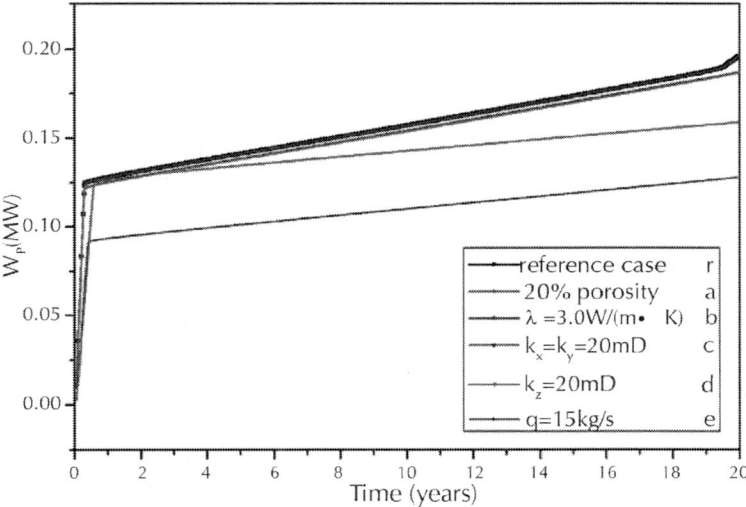

Figure 18: Sensitivity of pump power to various parameters.

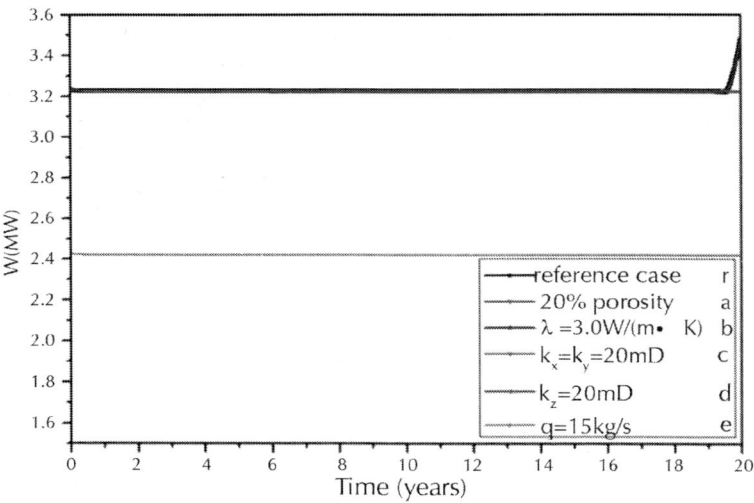

Figure 19: Sensitivity of electric power to various parameters.

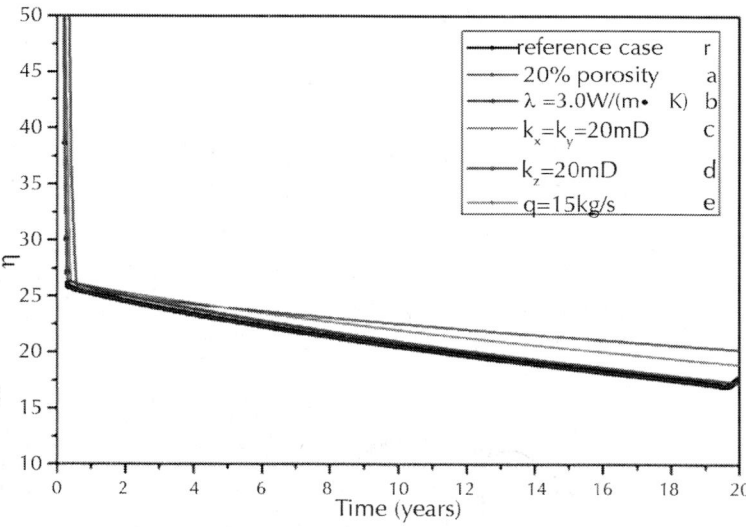

Figure 20: Sensitivity of energy efficiency to various parameters.

Sensitivity to φ

Previous EGS field tests prove porosity of fractured granite reservoir normally ranges within 1~7%[1] and [22]. Fan found the porosity of the deep fractured granite reservoir at Yangbajing geothermal field is about 10% [25]. Fig. 15, Fig. 16, Fig. 17, Fig. 18, Fig. 19 and Fig. 20a show the sensitivity of electricity generation to the reservoir porosity φ. Fig. 15 and Fig. 16a show that increasing from 10% to 20% eliminate the declining stage and there is only the stable stage during the 20 years, so the well only produces liquid water and h_{pro} basically retains unchanged. This is because this increment of φ results in an increment of P_{pro} from 8.76 to 3.58 MPa of the reference case to 8.69~4.78 MPa of the case a. Thus all through the 20 years P_{pro} is higher than 3.84 MPa, as shown in Fig. 17a. Fig. 18a shows that increasing φ from 10% to 20% results in a drop of W_p. This is according with equation (1), because a higher P_{pro} will decrease the pump height and reduce the pump power. As stated above, due to quite small variation of T_{pro}, h_{pro} basically retains unchanged during the stable stage. Fig. 19a shows that in case a h_{pro} basically retains unchanged during the entire 20 years, thus W also retains at a constant level of 3.23 MW according to equation (2) due to the elimination of the declining stage. Fig. 20a shows that in case a η is increased to the level of 50.00~20.29 from the level of 50.00~17.16 of the reference case, and meanwhile the efficiency rise during the declining stage is eliminated. This is according with equation (3), because W basically retains unchanged at 3.23 MW while W_p is decreased to 0.04~0.16 MW. On the whole, φ has a significant influence on the electricity generation performance of the single well system, and higher φ will be in favor of improving the electricity generation performance. This is because φ represents the resource amount of thermal water in the in-situ reservoir, a higher φ means there is more geothermal water in the reservoir provided for exploitation, and this slows down the decreasing of P_{pro} and makes favorable condition for heat production and electricity generation. This feature is similar to that in gas production performance using a single horizontal well [18], but is different from that in heat production using injection-production well system. In the injection-production system, the produced fluid is mainly from the injected cold fluid, normally there is little in-situ fluid in the reservoir, and the water production rate is determined by reservoir impedance and pressure difference between the injector and the producer, so the electricity generation performance has very limited relationship with φ [1], [22], [39], [49] and [50].

Sensitivity to λ

The heat conductivity λ of granite rock is closely related with structure and composition of the rock, is obviously influenced by temperature while very weakly affected by pressure; λ basically decreases with increasing temperature for the granite [7]. In this work the influence on the heat conductivity of temperature and pressure is neglected. Fig. 15, Fig. 16, Fig. 17, Fig. 18, Fig. 19 and Fig. 20b show that increasing λ from 2.5 W/(m·K) to 3.0 W/(m·K) has very limited influence on the heat production and electricity generation performance. This is because the in-situ temperature has only decreased by several Celsius degrees, as shown in Fig. 11, and the heat conduction effect is very small in the rock. In the reservoir the heat is transported mainly by convection, so the weak heat conduction will have very small impact on the system production performance. This feature is similar to that in the two horizontal wells system [22].

Sensitivity to $k_x = k_y$ and k_z

Fan pointed out that the intrinsic reservoir permeability ranges within 1~25 mD and maybe the horizontal intrinsic permeability is different from the vertical intrinsic permeability of the deep reservoir at Yangbajing geothermal field [25]. Fig. 15, Fig. 16, Fig. 17, Fig. 18, Fig. 19 and Fig. 20c and Fig. 15, Fig. 16, Fig. 17, Fig. 18, Fig. 19 and Fig. 20d show the sensitivity of electricity generation to the horizontal intrinsic permeability and to the vertical intrinsic permeability, respectively. Fig. 16c or d shows that increasing the intrinsic permeability from 10 mD to 20 mD will eliminate the declining stage and make the system only produce liquid water during the 20 years. This is because higher intrinsic permeability provides better flowing condition and makes the transportation toward the horizontal well easier. So higher intrinsic permeability slows down the decreasing of P_{pro}, as shown in Fig. 17c and d. Compared with the reference case, the minimum P_{pro} is increased to 3.86 MPa and 3.88 MPa, respectively, and both are higher than 3.84 MPa. So during the 20 years the system only produces liquid water and h_{pro} basically remains unchanged, as shown in Fig. 15c and d. Similar to case a, based on equation (2) W basically always remains at 3.23 MW, as shown in Fig. 19c or d. Fig. 18c or d shows that W_p slightly decreases with increased intrinsic permeability, this is because higher intrinsic permeability means higher P_{pro}. Consequently, based on equation (3) is slightly increased and the energy efficiency rise during the declining stage is also eliminated, as shown in Fig. 20c and d. Comparing case c with case d it can be readily found that the effect of horizontal intrinsic permeability variation is basically the same with that of vertical intrinsic permeability variation, and the heterogeneity of the fractured granite reservoir permeability within 1~25 mD has very limited impact on the electricity generation performance. On the whole, the intrinsic reservoir permeability has an important influence on the system electricity generation performance. Higher intrinsic permeability will improve the flowing condition, increase the P_{pro}, reduce the W_p and improve the system energy efficiency. These are similar to the conclusions in gas production performance through a single horizontal well or heat production performance through two horizontal wells [18] and [22].

Sensitivity to q

Fig. 15, Fig. 16, Fig. 17, Fig. 18, Fig. 19 and Fig. 20e show the sensitivity of electricity generation to the fluid production rate. Fig. 17e shows that decreasing q from 20 kg/s to 15 kg/s results in a drop of P_{pro} from the level of 8.76~3.58 MPa to 8.69~4.40 MPa, making P_{pro} always higher than 3.84 MPa. So during the 20 years the well only produces liquid water and h_{pro} basically retains unchanged, as shown inFig. 15 and Fig. 16, and the declining stage is eliminated. Fig. 18 and Fig. 19 show that decreasing q from 20 kg/s to 15 kg/s results in a drop of W_p from the level of 0.04~0.20 MW to 0.04~0.13 MW, also a drop of Wfrom 3.23 MW to 2.42 MW, and these are according with equations (1) and (2). This is because lower qslows down the decrease of P_{pro} and improves the P_{pro}, thus decreases W_p based on equation (1) and reduces W based on equation (2). Fig. 20e shows that decreasing q from 20 kg/s to 15 kg/s results in an increment of   from the level of 50.00~17.16 to 50.00~18.94. The energy efficiency is slightly increased and this is according with that in two horizontal wells system [22]. The vaporization of the liquid water in the reservoir will lead to the chemical deposition and reservoir blockage, and this may stop the operation of the system. In order to avoid the immature vaporization in the horizontal well, the water production rate must be lower than a limit which will control the decrease of P_{pro} during the whole production period; for economic performance of the electricity generation the fluid production rate should be higher than a limit. We can find that there is a trade-off between the rock and water chemical properties and the thermodynamic properties of water, which will finally determine the optimized fluid production rate at the deep reservoir [1],[22] and [41]. This is similar to that in the traditional exploitation of hot dry rock resources [42] and [44]. On the whole, the fluid production rate has a significant influence on the electricity generation performance. A reasonable fluid production rate will restrain the immature vaporization in the wellbore, improve the stability of fluid flow, and increase the system energy efficiency.

CONCLUSIONS

In this work, we numerically investigated the electricity generation potential from the 950~1350 m fractured granite reservoir through a single horizontal well near well ZK4001 at Yangbajing geothermal field over a period of 20 years. According to the study results, the following conclusions are made.

- The entire electricity production process can be divided into two stages: the stable stage and the declining stage. In the reference case, during the 20 years the electric power mainly ranges within 3.23~3.48 MW and the energy efficiency mainly ranges within 50.00~17.16.

- During the stable stage, the well only produces liquid water and the production specific enthalpy basically retains unchanged; during the declining stage, the well produces the mixture of liquid and steam, the production temperature slightly declines and the production specific enthalpy gradually increases.

- During the 20 years the bottomhole pressure gradually declines. If the bottomhole pressure declines to lower than the saturated vapor pressure, the liquid will vaporize in the wellbore.

- During the 20 years the pump power gradually increases. During the declining stage when the well produces the mixture of liquid and steam, the pump power rises more quickly.

- During the stable stage, the electric power retains unchanged; in the declining stage, the electric power slightly increases due to the latent heat of vaporization of water.

- During the stable stage, the energy efficiency gradually declines; in the declining stage, the energy efficiency slightly rises. If there is no declining stage, the energy efficiency will decline throughout the production period.

- The heat recovery factor is very small because there is no cold water injected into the reservoir to extract the heat energy in the rock.

- The reservoir temperature only decreases slightly, this is because the local heat energy is redistributed as a result of the vaporization. The pressure at the top reservoir is minimum, the vaporization first starts here and the temperature at the top reservoir first begins to decline. The temperature interfaces are slightly inclined and migrate downward as heat production goes on.

- The pressure in the upper reservoir retains unchanged at the saturated vapor pressure once the pressure declines to lower than the saturated vapor pressure. The pressure interfaces are slightly inclined and migrate downward. In the lower reservoir the pressure gradually declines.

- The steam saturation interfaces are inclined and migrate downward. The upper two-phase liquid steam zone gradually expands, standing for that the liquid gradually vaporize to steam in the lower liquid zone.

- Analysis of sensitivity to φ indicates that φ has a significant influence on the electricity generation performance of the single well system. Higher φ means more in-situ fluid in the reservoir, this increases the bottomhole pressure and decreases the pump power, and finally improves the energy efficiency.

- Analysis of sensitivity to λ indicates that λ has very limited impact on the electricity generation performance of the single well system. This is because in the reservoir the heat is transported mainly by convection and the heat conduction is very weak.

- Analysis of sensitivity to $k_x = k_y$ and k_z shows that intrinsic reservoir permeability has an important influence on the electricity generation performance. Higher intrinsic reservoir permeability will improve the flow condition, increase the bottomhole pressure, decrease the pump power and improve the energy efficiency; furthermore higher intrinsic reservoir permeability is favorable for eliminating the vaporization process in the wellbore.

- Analysis of sensitivity to q shows that q has a significant influence on the electricity generation performance. A reasonable fluid production rate will restrain the immature vaporization in the wellbore, improve the stability of fluid flow, and increase the system energy efficiency.

According to the commercial criterion of enhanced geothermal system, the electric power and energy efficiency of the single horizontal well system are acceptable. Main parameters that determine the electricity generation performance are the reservoir porosity, the intrinsic reservoir permeability and the fluid production rate. The electricity generation performance of the system will be more optimized when all the parameters are under more reasonable conditions, such as higher reservoir porosity or higher intrinsic permeability or more appropriate fluid production rate.

However, because of lacking for deep geological data of the fractured granite reservoir, in this work we have made some assumptions and simplifications. Although similar assumptions have been adopted in previous studies, the effect of these approximations on the electricity generation performance needs further research in the future.

ACKNOWLEDGMENTS

The authors gratefully appreciate the financial support from the National High Technology Research and Development Program of China (Grant 2012AA052802); the Director Fund Projects of the Guangzhou Institute of Energy Conversion, Chinese Academy of Sciences (Grant y107a41001); the Science and Technology Innovation Special for Graduate Student, Chinese Academy of Sciences (y207y81001).

REFERENCES

1. Tester JW, Livesay B, Anderson BJ, Moore MC, Bathchelor AS, Nichols K, et al. The future of geothermal energy: impact of enhanced geothermal systems (EGS) on the United States in the 21st century. An assessment by an MIT-led interdisciplinary panel; 2006.

2. Kuriyagawa M, Tenma N. Development of hot dry rock

technology at Hijiori test site: program for a long-term circulation test. Geothermics 1999; 28(4e5): 627e36.

3.   Duchane DV. Hot dry rock: a realistic energy option. Geotherm Resour Counc Bull 1990; 19(3):83e8.

4.   Evans K. Enhanced/engineered geothermal system: an introduction with overviews of deep systems built and circulated to date. In: China geothermal development forum. Beijing Sep. 13, 2010. pp. 395e418.

5.   Baria R, Baumgärtner J, Rummel F, Pine RJ, Sato Y. HDR/HWR reservoir: concepts, understanding and creation. Geothermics 1999; 28:533e52.

6.   Xiulan Yin. Unlimited prospect for utilization for hot dry rock geothermal resources in Chinese. http://www.cn-mineralwater.com/show.aspx? id¼1267&cid¼11; 2010.

7.   Zhao YS, Wan ZJ, Kang JR. Introduction to HDR geothermal development. Beijing: Science Press; 2004 in Chinese..

8.   Xu T, Zhang Y, Zeng Z, Bao X. Technology progress in an enhanced geothermal system (hot dry rock). Sci Technol Rev 2012; 30(32):42e5 in Chinese..

9.   Mao H, Ma H. Energy Eng 2010; 5:25e8 in Chinese..

10.   Wan Z, Zhao Y, Kang J. Forecast and evaluation of hot dry rock geothermal resource in China. Renew Energy 2005; 30:1831e46.

11.   Wang J, Hu S, Pang Z, He L, Zhao P, Zhu C, et al. Estimate of geothermal resources potential for hot dry rock in the continental area of China. Sci Technol Rev 2012;30(32):25e31 in Chinese..

12.   Zhou A, Zhao Y, Guo J, Zhang N. Study of geothermal extraction scheme of hot dry rock in Tibetan Yangbajing Region. Chin J Rock Mech Eng 2010; 29(Supp. 2):4089e95 in Chinese..

13.   Zhao Y, Feng Z, Xi B, Zhao J, Wan Z, Zhou A. Prospect of HDR geothermal energy exploitation in Yangbajing, Tibet, China, and experimental investigation of granite under high

temperature and high pressure. J Rock Mech Geotech Eng 2011;3(3):260e9.

14. Feng Z, Zhao Y, Zhou A, Zhang N. Development program of hot dry rock geothermal resource in the Yangbajing Basin of China. Renew Energy 2012; 39:490e5.

15. Bruno MS, Serajian V, White KLN, Elkhoury JE, Detwiler RL. Advanced horizontal well recirculation systems for geothermal energy recovery in sedimentary formations. Geothermal Resources Council; 2012.

16. Polsky Y, Capuano L, Finger J, Huh M, Knudsen S, Chipmansure AJ, et al. Enhanced geothermal systems (EGS) well construction technology evaluation report. Sandia National Laboratories; 2008.

17. Li G, Moridis GJ, Zhang K, Li X. The use of huff and puff method in a single horizontal well in gas production from marine gas hydrate deposits in the Shenhu Area of South China Sea. J Pet Sci Eng 2011; 77:49e68.

18. Li XS, Li B, Li G, Yang B. Numerical simulation of gas production potential from permafrost hydrate deposits by huff and puff method in a single horizontal well in Qilian Mountain, Qinghai province. Energy 2012; 40:59e75.

19. Li XS, Yang B, Li G, Li B. Numerical simulation of gas production from natural gas hydrate using a single horizontal well by depressurization in Qilian Mountain Permafrost. Ind Eng Chem Res 2012; 51:4424e32.

20. Li G, Li XS, Zhang KN, Moridis GJ. Numerical simulation of gas production from hydrate accumulations using a single horizontal well in Shenhu Area, South China Sea. Chin J Geophys 2011; 54(9):2325e37 in Chinese.

21. Li G, Moridis GJ, Zhang KN, Li XS. Evaluation of gas production potential from marine gas hydrate deposits in Shenhu area of South China Sea. Energy Fuels 2010; 24:6018e33.

22. Zeng YC, Su Z, Wu NY. Numerical simulation of heat production potential from hot dry rock by water circulating through two horizontal wells at Desert Peak geothermal field.

Energy 2013; 56:92e107.

23. Dor J, Zhao P. Characteristics and genesis of the Yangbajing geothermal field, Tibet. In: Proceedings of the World Geothermal Congress 2000, KyushuTohoku, Japan, and 1083e8.

24. Introduction to geothermal fields in China in Chinese. http://wenku.baidu. com/view/4b4a9b1ba76e58fafab003b2.html; 2013.

25. Fan X. Conceptual model and assessment of the Yangbajing geothermal field, Tibet, China. In: Geothermal training programme, Orkusrofnun, Grensásvegur 9, IS-108 Reykjavík, and Iceland. Number 5 2002.

26. Zhao P, Dor J, Jin J. A new geochemical model of the Yangbajing geothermal field, Tibet. In: Proceedings of the World Geothermal Congress 2000 2007 2012. Kyushu-Tohoku, Japan.

27. The Landau geothermal power planthttp://www.bestec-for-nature.com/j_ bestec/index.php/en/projects/landau; 2013.

28. Sanyal SK, Butler SJ. An analysis of power generation prospects from enhanced geothermal systems. In: Proceedings of world geothermal congress 2005, Antalya, Turkey 2005.

29. Rutqvist J, Oldenburg CM. Analysis of injection-induced micro-earthquakes in a geothermal steam reservoir, the Geysers Geothermal Field, California. Lawrence Berkeley National Laboratory; 2008. 2013, http://escholarship.org/ uc/item/4qg7v6qz.

30. Shaik AR, Rahman SS, Tran NH, Tran T. Numerical simulation of fluid-rock coupling heat transfer in naturally fractured geothermal system. Appl Therm Eng 2011; 31:1600e6.

31. Pruess K. Modelling of geothermal reservoirs: fundamental processes, computer simulation, and field applications. In: Proc. 10th New Zealand geothermal workshop 1988.

32. Pashkevich RI, Taskin VV. Numerical simulation of exploitation of supercritical enhanced geothermal System. In: Proceedings of thirty-fourth workshop on Geothermal

Reservoir Engineering. Stanford, California: Stanford University; February 9e11, 2009.

33. Fakcharoenphol P, Hu L, Wu YS. Fully-implicit flow and geomechanics model: application for enhanced geothermal reservoir simulations. In: Proceedings in thirty-seventh workshop on geothermal reservoir engineering. Stanford, California: Stanford University; 2012.

34. Radilla G, Sausse J, Sanjuan B, Fourar M. Interpreting tracer tests in the enhanced geothermal systems (EGS) of Soultz-sous-Forêts using the equivalent stratified medium approach. Geothermics 2012; 44:43e51.

35. Birdsell S, Robinson B. A three-dimensional model of fluid, heat, and tracer transport in the Fenton Hill hot dry rock reservoir. In: Proceedings in thirteenth workshop on geothermal reservoir engineering. Stanford, California: Stanford University; January 19e21, 1988.

36. Pruess K, Wu YS. A semi-analytical method for heat sweep calculations in fractured reservoirs. In: Proceedings in thirteenth workshop on geothermal reservoir engineering. Stanford, California: Stanford University; January 19- 21, 1988.

37. McDermott CI, Randriamanjatosoa ARL, Tenzer H, Kolditz O. Simulation of heat extraction from crystalline rocks: the influence of coupled processes on differential reservoir cooling. Geothermics 2006; 35:321e44.

38. Watanabe N, Wang W, Mcdermott C, Taniguchi T, Kolditz O. Uncertainty analysis of thermo-hydro-mechanical coupled processes in heterogeneous porous media. Comput Mech 2010; 45:263e80.

39. Murphy H, Brown D, Jung R, Matsunaga I, Parker R. Hydraulics and well testing of engineered geothermal reservoirs. Geothermics 1999; 28:491e 506.

40. Wagner W, Kretzschmar HJ. International steam tables-properties of water and steam based on the industrial formulation IAPWS-IF97. 2nd ed. Springer; 2007.

41. Zhou DJ. Operation, problems and countermeasures of Yangbajing geothermal power station in Tibet. Electr Power Constr 2003; 24(10). 1e3þ9 in Chinese..

42. Pruess K. On production behavior of enhanced geothermal systems with CO2 as working fluid. Energy Convers Manag 2008; 49:1446e54.

43. Davis AP, Michaelides EE. Geothermal power production from abandoned oil wells. Energy 2009; 34:866e72.

44. Pruess K. Enhanced geothermal system (EGS) using CO2 as working fluid-A novel approach for generating renewable energy with simultaneous sequestration of carbon. Geothermics 2006; 35:351e67.

45. Zhang TH, Huang QZ. Pollution of geothermal waste water produced by Tibet Yangbajing geothermal power station. Acta Scientiae Circumstantiae 1997; 17(2):252e5 in Chinese..

46. Pruess K, Oldenburg C, Moridis G. TOUGH2 user's guide, version 2.0. Berkeley, CA, USA: Lawrence Berkeley National Laboratory; 1999.

47. Borgia A, Pruess K, Kneafsey TJ, Oldenburg CM, Pan L. Numerical simulation of salt precipitation in the fractures of a CO2-enhanced geothermal system. Geothermics 2012; 44:13e22.

48. Zeng YC, Su Z, Wu NY, Wang XX. Numerical simulation of deep geothermal energy mining by two vertical wells at Desert Peak field, USA. Mining Metall Eng 2013; 33(2). 8e13þ17 in Chinese

49. Brown D, DuTeaux R, Kruger P, Swenson D, Yamaguchi T. Fluid circulation and heat extraction from engineered geothermal reservoirs. Geothermics 1999; 28:553e72.

50. Zeng YC, Wu NY, Su Z, Wang XX, Hu J. Numerical simulation of heat production potential from hot dry rock by water circulating through a novel single vertical fracture at Desert Peak geothermal field. Energy 2013; 63: 268e82.

Citations

CHAPTER 1

A. Cunha, M. Pacheco and J. Bergmann, "Influence of the Chemical Composition of Completion Fluids on the Propagation of Electromagnetic Waves within Oil Wells," Engineering, Vol. 4 No. 12A, 2012, pp. 966-971. doi:10.4236/eng.2012.412A122.

CHAPTER 2

Jing-Chun Feng, Yi Wang, Xiao-Sen Li, Gang Li, Yu Zhang, Zhao-Yang Chen, Effect of horizontal and vertical well patterns on methane hydrate dissociation behaviors in pilot-scale hydrate simulator,

Applied Energy, Volume 145, 1 May 2015, Pages 69-79, ISSN 0306-2619, http://dx.doi.org/10.1016/j.apenergy.2015.01.137.

CHAPTER 3

J. Parga, G. Munive, J. Valenzuela, V. Vazquez and G. Zamarripa, "Copper Recovery from Barren Cyanide Solution by Using Electrocoagulation Iron Process," Advances in Chemical Engineering and Science, Vol. 3 No. 2, 2013, pp. 150-156. doi: 10.4236/aces.2013.32018.

CHAPTER 4

Majid Ali Abbasi, Daniel Obinna Ezulike, Hassan Dehghanpour, Robert V. Hawkes, A comparative study of flowback rate and pressure transient behavior in multifractured horizontal wells completed in tight gas and oil reservoirs, Journal of Natural Gas Science and Engineering, Volume 17, March 2014, Pages 82-93, ISSN 1875-5100. http://dx.doi.org/10.1016/j.jngse.2013.12.007.

CHAPTER 5

Supalak Parn-anurak, Thomas W. Engler, Modeling of fluid filtration and near-wellbore damage along a horizontal well, Journal of Petroleum Science and Engineering, Volume 46, Issue 3, 15 March 2005, Pages 149-160, ISSN 0920-4105, http://dx.doi.org/10.1016/j.petrol.2004.12.003.

CHAPTER 6

M. Khajenoori, A. Haghighi-Asl, J. Safdari, M.H. Mallah, Prediction of drop size distribution in a horizontal pulsed plate extraction column, Chemical Engineering and Processing: Process Intensification, Volume 92, June 2015, Pages 25-32, ISSN 0255-2701, http://dx.doi.org/10.1016/j.cep.2015.03.021.

CHAPTER 7

Reza Ettehadi Osgouei, A. Murat Ozbayoglu, Evren M. Ozbayoglu, Ertan Yuksel, Aydın Eresen, Pressure drop estimation in horizontal annuli for liquid–gas 2 phase flow: Comparison of mechanistic models and computational intelligence techniques, Computers & Fluids, Volume 112, 2 May 2015, Pages 108-115, ISSN 0045-7930. http://dx.doi.org/10.1016/j.compfluid.2014.11.003.

CHAPTER 8

N. Hadia, L. Chaudhari, Sushanta K. Mitra, M. Vinjamur, R. Singh, Experimental investigation of use of horizontal wells in waterflooding, Journal of Petroleum Science and Engineering, Volume 56, Issue 4, April 2007, Pages 303-310, ISSN 0920-4105, http://dx.doi.org/10.1016/j.petrol.2006.10.004.

CHAPTER 9

Yu-Chao Zeng, Neng-You Wu, Zheng Su, Jian Hu, Numerical simulation of electricity generation potential from fractured granite reservoir through a single horizontal well at Yangbajing geothermal field, Energy, Volume 65, 1 February 2014, Pages 472-487, ISSN 0360-5442, http://dx.doi.org/10.1016/j.energy.2013.10.084.

Index